TANS

I0140629

THE TANS COLLECTION

VOLUME IV

You Called. We Went.

Edited by Wayne Munkel
& John Klawitter

TANS
The TANS Collection, Volume IV

Copyright 2015 by the Southeast Asia ASA Association
David Ulm Contact Person

Cover Photo courtesy of George Murphy
Original Cover Concept by Wayne *Lucky* Rupp
Cover Illustration by MB10, Marquette, MI State Prison
Proofreader Terry Werner
Chief Editor Wayne Munkel
Editor John *Mad Jack* Klawitter

ISBN: 1-938674-10-3

DoubleSpin
DoubleSpin Publishing
A Division of Dancing Bear Ent. LLC
Printed in the United States of America
Published 05 March 2016

Dedicated to all ASA Veterans
Thank you for your service and sacrifice

CONTENTS

Chapter 1
In the Beginning

Chapter 2
Everyday ASA

Chapter 3
Special Events

Chapter 4
Time and Place is Everything

Chapter 5
ASA in Other Places

Chapter 6
Reflections on Vietnam

Chapter 1
In the Beginning

Beginnings
By Robert Flanagan

A few days before Christmas, 1963, I handed in my final paper, breathed a deep sigh of relief and drove home to South Milwaukee in a light snow, having just completed a two year assignment as a college student under the U.S. Army's kindly sponsorship. Earlier, in 1962, winding down two years in Asmara, I'd secured a slot in a highly contested education program wherein the Army sent select personnel to college to study specific subjects, in my case Electrical Engineering. Kennedy had said we were going to the moon; we needed to step up our collective technical capabilities, and though the president had gone on ahead of us from Dallas, his emphasis on science, math, and engineering was still the drumbeat marched to in Washington.

Skipping here all the boredom of studying a subject I'd no interest in and no calling for, I did manage to cram in a passel of hard-won semester hours at Milwaukee School of Engineering (MSOE). If nothing else, the school assignment had pre-empted orders for my return from Africa to CONUS, to the 317 ASA Bn at Fort Bragg. I'd not looked forward to the airborne aspects of that assignment, being at the time an old married man of twenty-six with two sloe-eyed children who doted upon a live father; and besides, been there, done that. Made one whole jump under circumstances designed to… well, let's save that for another story, another time.

So I'd gone to school at MSOE carrying a slide rule, rather than at Bragg dragging a T-5 chute, and now I had used up all the assignment time I could cadge out of that windfall. For the Army program, I'd been released by ASA to DA, with student status in Milwaukee and reporting to

Student Detachment, HQ, Fifth Army, in Chicago, 90 miles south. My assigned duty while in school: 1) go to class faithfully and 2) study rigorously. I lived on the economy with my family, as there was no military housing in Mill Town; I wore civilian clothes; I made no formations, no roll calls, no PT ... and pay came by monthly check in the mail. What's to argue?

When school was finished, Fifth Army, not knowing what to do with me, reassigned me to ASA. But these things take time. Fifth Army had me report to a U.S. Army Reserve Unit in South Milwaukee, where I would run errands, drive vehicles, type correspondence and training curricula and otherwise provide grunge labor while awaiting orders. Nice group of guys on the staff of that reserve infantry battalion: a light Colonel, a major, two captains and five or six senior sergeants. Need I say in joining that fraternity, labor was not intensive, even for an E-5.

On 2 Mar 64, the Colonel called me in and handed me a teletype tear sheet—my orders: Asg to: 3d Radio *Recovery* Unit (PROV) (6082), APO 143, San Francisco, Calif. What? I never did discover the provenance of Recovery. But more to the point, where was APO 143? The senior clerk made a call to Fifth Army, and stared up at me with a curious gaze. 'Saigon,' he said, I thought a bit ominously. Duhh, where's Saigon? I'd read Graham Greene and knew the name, but I hadn't a clue. I walked over to the Colonel's Brown-Mule-Chewing-Tobacco-barn-wall-sized map of the world, and with his and a couple of stray malingerers' help, found Saigon. Waaay-y-y out there, over in Southeast Asia on the periphery of a land called Indochina.

"Hey, gotcha covered," someone said and fetched an SFC who had recently returned from a tour with MAAG-Vietnam; he came in and set us straight. It was no longer Indochina, a former French conglomerate of colonies, but was now four independent countries: South and North Vietnam, Laos and Cambodia. The SFC said with a bit of a pout, "'s been in the papers, lately. But ..." Following his pointing finger down to the map legend, I saw the date 1952 displayed prominently. The Colonel mumbled something about "... updating the map."

Now that I had a focus, I bent the ear of that MAAG Sergeant and that, in the final analysis, was probably not the wisest course. He was a bitter man because of some perceived maltreatment by his ARVN counterparts, and he was full of tales of dark deeds and misdeeds, malfunctions, and military/political malingering. He also knew nothing of ASA personnel in 'Nam. But I was returning to duty as a 058.30; I knew ASA Ops had to be in my future.

When I'd packed my family south to Mississippi and had set them up, I departed with a sense of unease. Things did not auger well for the forthcoming assignment. My wife wisely insisted that the SFC was only one view; I should wait and form my own opinions. At the same time, though she was right and I eventually settled into acceptance and a come-what-may attitude. For her part, she was uneasy and, with the three children, cried upon my departure.

Arlington Hall was not forthcoming with additional information, revealing only a port call, place and date at Travis AFB, Calif. Arriving there mid-March in the middle of the night, I joined a gaggle of other military of all four major services, among them a small cotillion of ASA troops whom I identified by their First Army patches and collar brass, all fresh out of Fort Devens. Clustered in a small group, they looked upon my SP-5 rank as promising leadership and military wisdom. I disavowed them of that fairy tale by invoking my new-found civilian status (not a real state-of-being, but an attitude engendered by my two years in civilian clothes without military oversight). But we boarded the C-135 in a group and managed to stay together across the Pacific to Hawaii, Wake Island, a weather diversion to Okinawa, on to Clark AFB in the PI, and eventually Tan Son Nhut.

You've read of, or experienced, that ubiquitous arrival many times over. After landing and a 47-mile taxi, the bird rocks to a halt, the door pops open and all the dismal atmosphere of three hells spills in the door, heat leading the assault, at one with humidity, and mind-numbing stench hard on their heels. The babble of indecipherable tongues would follow later, as well as physical threats, mosquitoes with their bounty of malaria and dengue fever, venomous snakes, scorpions and a billion unidentifiable other risks to life and limb. You know the scene; we need not re-visit. Oh, and there was also the enemy: Charlie!

Upon our only slightly unique arrival, up the ladder steps bounded an Army Major wearing Signal Corps brass. He stopped the first disembarking troops and called out down the length of the plane, "AIS personnel keep your seats; all other military personnel deplane Asap."

He plopped into an open seat and waited while curious troops shuffled by him. When all were gone except the group of ASA-ers and a couple more shifty-looking characters, the Major indicated to the Air Force coolie to shut the door. He rose and faced us, as we gathered near. "People, remove your AIS collar brass." He proffered a large manila courier envelope and said, "Place all the brass in here, along with any watch fobs, ballpoint pens, pictures, notes, business cards ... *anything* with an AIS or the old ASA logo on them."

[This acronym, AIS, represented one more illogical merging/rending-asunder of the intelligence world. When I changed services, leaving the Marines and enlisting in the Army in 1960, it was into ASA. Sometime in the interim, ASA had merged with Military Intelligence to form Army Intelligence and Security (AIS); but never fear, it was to be later changed back to something near its origins ... until such time as ASA went away in 1976 and Intelligence and Security Command (INSCOM) came into being—an AIS look-alike.]

"Sir," asked a puzzled PFC. "Will we get our gear back?"

"Unlikely, soldier. I'm not gonna shit you; this stuff will go into a safe somewhere and will be lost track of before you're finished with your tour, which would be the only time you could claim it back. No, it's a goner."

"Sir," pushed the PFC, "what the hell's—"

"OK, troop. Here's the deal. All members of AIS, primarily ASA-ers, have a bounty on their head in 'Nam. The VC—the Viet Cong, the bad guys ... Charlie—recognizes the significance of your presence here, and since you constitute a far greater individual threat to him than any of the other scattering of troops in country—except possibly the Green Beanies—consisting mostly of MAAG, logistics people, helicopter drivers and what-all, he's particularly anxious to blow your shit away. And he's willing to pay for information identifying you among all the drones," he said in a condescending manner. "Don't make it easy for him."

Well, that pretty much set the tone for that day's dance. We did as ordered and went, collar-naked to Davis Station, where all but the couple of shady strangers were to be billeted. We drew Signal Corps collar brass upon sign in, for those rare occasions when we'd be in dress uniform. In briefings we learned quickly of the recent deaths of two 3rd RRU soldiers, killed at a softball game, blown up by an explosive charge planted under their bleacher seats. We heard of the earlier downtown movie theater bombing that killed and wounded Americans. And we got the party line on Tom Davis and why Davis Station was so named.

But this was just the beginning. Over the following twelve-month lifetime I would experience a number of disquieting events— including the interception and killing of the late night Viet Cong sapper who shattered the tranquility of the evening movie and was killed by a spooked perimeter guard, when the subject VC turned out to be a water buffalo crashing about in a cane break behind the motor pool; the series of coups and counter-coups as one general after another sought to take the place of power of the recently assassinated President Diem; loss of an aircraft and near-loss of the pilot, Ray McNew and Op, Dan Bonfield; an explosion in the civilian air terminal at TSN wounding three ASA-ers (2LT Chladek, SSGs Moore and Lenihan) and other troops, and many other curious happenings—but none, I think, ever rose to the level of sheer uneasiness engendered by that major's reception speech. Unless it was when I began flying ARDF in those tinker toy aircraft.

Mavis Manymoney
By John *Mad Jack* Klawitter

Autumn 1962: Mavis Manymoney was some kind of witch. At least that's what she told people in our apartment complex. It was the natural thing, she said, the way they lived on the island of Jamaica, and her Mom and Granny had taught her everything they knew about the mumbo and the juju, the incantations, the stick-pin dolls and the rest of it. We lived in the same complex, a U-shaped three story building with parking underneath. This was over on Strathmore in the Village a few blocks off the UCLA campus, and of course that sort of gossipy thing spread around pretty fast. She'd give you a predatory grin and ask for a fingernail clipping or if she could borrow your comb, just a few strands of hair was all she needed, and the way she said it you couldn't be sure she was joking around or not. She said she kept other witches out with a trail of blessed sand spread across her doorway at night, and anybody walking past her place to their own could see the little trail of beach sand, spread out in a thin wiggly line, right there. With her hair-trigger temper and that grim faced stare of hers, folks in the complex tended to give her plenty of space to go on and be as crazy as she wanted.

Unfortunately, I had been on her bad side since the first time we'd met. I'd been up on the rooftop deck practicing fencing with an undergrad I was sweet on. Not that I knew that much about fencing, but she was trying to make the UCLA fencing team, and any old body would do. With my customary grind-it-out tenacity, I had gone into a personal training regimen and after a while I'd improved at it to where I had to let her stick me every now and then so she wouldn't get too pissed off.

We were in the middle of one of those longish *six et quatre et six et quatre* bouts, rattling back and forth across the flat rooftop while the morning sun began to burn through the low hanging fog off the Pacific. Suddenly the door to the stairway banged open and we were confronted *mid-quatre* by a screaming dervish with a thick layer of blue goop smeared all over her face, bright yellow crap on her lips and pink plastic curlers in her hair. This fierce apparition was wearing only an open bathrobe that showed off plenty of her ebony black skin. She had a thin but not unattractive body, and from what little I could make out of her face, high cheekbones, a thin nose and a thin-set mouth made for frowning and disapproval. All in all, it was an astounding and unexpected entrance. I took off my mask, gaping at the semi-naked spectacle.

The newcomer screamed at us, waving her fists in the air and oblivious to the fact that she was charging at two people with fencing foils in their hands who could skewer her like a barbecued chicken.

She yelled, "What are you crazy moron white peoples doing, stomping on the ceiling over my head in the middle of the night?"

My partner sighed, "Mavis, it is ten in the morning. And be reasonable, this is a public deck."

"You got no right to be stomping on my roof, not no time of day, bitch!"

It looked to be the beginning of World War III, but just as my partner pulled off her mask and took a firmer grip on her foil, Mavis held up one hand like she was stopping traffic. She stared as if seeing me the first time, looking me up and down like I was a popsicle treat or maybe an Almond Joy candy bar.

"'Nilla Boy!" She purred, moving toward me like a big black cat. She reached out and stroked my cheek. "Sweet, delicioso 'Nilla Boy! I been lookin' to meet you. Aww, now, here now, you skin so milky white and pretty…"

I took an involuntary step back, and the moment was interrupted as five of Mavis's house guests in various states of undress came boiling up from the floor below and spilled out onto the deck. Mavis raised a hand in their direction. "No, back off, islander-comrades, I got this in control. We just about done here." She paused and then dramatically pointed at me, "'Cepting I ain't done with you, 'Nilla Boy. Fact is, we, you-an-me, we just about to begin!" And with a sweeping flourish she gathered her orange robe about her hips in a queenly fashion and in another few seconds was gone.

A week went by before I saw her again. I was on campus, on my way down the sidewalk steps that led from the Grad English Department past Pauley Pavilion.

"Hey, 'Nilla Boy, why so glum in the face, mon?" She was sitting on a bench next to a small cluster of student activity sign up booths. In a tight blouse and short shorts, she was hardly recognizable as the fierce apparition that had interrupted my fencing session.

Feeling like a seed in the wind, I drifted over to sit next to her. "I've just been to see my advisor."

"What, you flunking out?"

"No, just the opposite. He wants to fast track me for a doctorate."

"Whoa, whoa, whoa, the man hands you the world on a silver platter and you goin' all glum?"

"It's not what I want. I came here to learn how to write, not how to catalog things."

"Well, 'Nilla Boy, you do what the man wants, he gives you what you want."

"Not this time." I eyed her, liking the way her nipples showed through her thin blouse. "Don't you wear any underwear?"

"You always say what you think?"

"Usually. Don't you?"

"Course not. But that ain't no way to talk to a lady."

I smiled, thinking but not daring to say the words *loose lady*, "Sorry. What are you doing here? Going to sign up for cheerleading?"

"Nooo, the opposite. I'm here to tear things down."

"What things?"

"Everything. All of this. Do you know my friend, here?" She put one arm through that of a middle aged man sitting next to her. The fellow was chunky and balding, and reminded me of an accountant or some middle management flunky.

He waved a casual hand in my direction, "Gus. Gus Hall," he said. He turned to Mavis. "Bring him along tonight. We like disillusioned."

As I watched the man slowly walk away, my own mind was already back on my troubles. My professor/advisor, who pretty much owned my scholastic destiny, had just offered me a fast track to earn my doctor's degree in English Lit. All I had to do was spend two or three years buried in the library stacks, researching possible connections between two English cleric poets. And I had turned him down.

"You're a fool, young Master Klawitter," the prof said. He arched his eyebrows and looked down his nose at me. "You're throwing away your golden ticket to a life of academic ease and accomplishment."

"But I'm not interested in minor English poets from two or three centuries ago."

"Nobody cares what you want, Master Klawitter. Take it or take a hike."

In the pleasant outdoors, Mavis was giving me an odd look. "You don't even know who that was you just met, do you, 'Nilla Boy?"

"Sure. Gus Bell. He said so."

"Gus *Hall*," she corrected me. The Gus Hall. Head of the CPUSA."

"What's the CPUSA?" I asked.

I went anyway, and the meeting turned out to be a dismal flop. About fifty students turned out for a rally outside Raleigh Hall. There were a few posters on sticks with slogans that I didn't understand.

It didn't help that I was a half hour late and Gus had come and left by the time I got there. Mavis furiously shoved a stack of pamphlets in my face, "Here. Hand these out."

"What's wrong?"

"We arranged for a TV crew to show up. Channel 5. We paid the rotten bastards."

"You paid for a film crew to show up?"

She looked at me like I was a moron. "How else do you get them here?"

"Well, apparently it didn't work."

For a moment she looked like she might attack me physically, but something else distracted her and she huffed away. The night air off the ocean had me shivering. I should have worn a warmer jacket. I found a spot out of the cold and examined the flyers she had given me. The headline read "Invader-Gangster American Colonialists—Get out of Vietnam!" There was a picture of what looked like a World War II tank rearing up to crush some startled cone-hatted peasants who were working in a rice field.

Later that night, while sharing a big bottle of red wine, Mavis lectured me on the deficiencies in my upbringing and education.

"You a know-nothing white boy from the house of privilege."

"And you?"

"Don't you start in on me, boss-man! Down in Kingston, me only 16, I see the policeman flunky of the colonials blow away my sister, one round pistol shot catch her in the eye."

"Yes, but I wasn't 'starting in on you', as you say."

"Oh, yeah, that's what you say so, yeah. It always be the same. You here because you want some of this hot black mamma. That what you want, 'Nilla Boy!"

Her one-room apartment generally was teeming with riffraff, but for the moment it was deserted. She crawled across the pillow strewn floor and launched herself at me.

"Well, here I am!" We made love like wild animals; at least she did, while I did my best to keep up. And over the next few days when we weren't going at each other, we argued like bitter enemies about the flaws of Western Civilization, the greed and corruptness of the Democratic society, and the brighter better way that was coming.

"You don't know what the hell you are talking about," I shouted. "You've never been to a communist country, and you sure as hell have never been to Vietnam!"

"I don't need glasses to see black-and-white right in front of my face!"

I made the excuse that I was late for class, and stormed out like I usually did.

On my way over to campus a friend of my fencing partner fell in step alongside me. "What, you give up on fencing lessons?"

"I don't know. That wasn't going anywhere fast."

"Hope you're using condoms with the black witch."

"Well, sure, of course. Why do you ask?"

"She's on record saying she wants a white baby, *any white baby.*"

"So?"

"Well, you know Andy was hanging tight with her over the summer. He found somebody had punched needle holes in his Trojans. She hasn't used your bathroom, has she?"

I said she hadn't, except for a time or two. I still had that lingering worry she was going to get a few hairs from my brush and do her island spells on me. But later that day I did a condom check and there was a neat little hole through the plastic wrapper on every one of my packs. They were nearly invisible and I had to look really close to be sure. So right there I was feeling a sudden incentive to get out of town.

While I sometime told myself (and later, literary agents and film producers) that I'd signed up for the Army to see war for myself like my heroes Hemingway and Crane (at the time I didn't know Crane had never been to war) and maybe Heller and Mailer, the truth was clearly not so black-and-white. As the months passed, my concern faded over whether I'd sired a child by the witchy Jamaican. I was sure that, if I had given her the little light brown baby of her dreams, she would have done some sort of tracking spell and pursued the father to the far corners of the world. On the other hand, considering how things turned out, there were plenty of times I was convinced she had tossed some sort of rotten voodoo luck in my direction just for spite. *Gus Hall is a real person. I don't care if he tries to deny this story. (if he's still alive) Manymoney is not Mavis's real name. Her, I'm still afraid of.

The Ant & The Elephant
By John *Mad Jack* Klawitter

John F. Kennedy was the best motivational speaker I'd ever heard. I actually never saw him in person, just black-and-white on the TV screen, but I thought he was better than Norman Vincent Peale and Martin Luther King and Everett Dirksen all stuck together, and when he started talking about the moral imperative to fight the good brushfire wars and save the Lao, the Vietnamese, the Cambodians and the Thais from falling to the Communist Menace, Americans cheered and signed up for the armed forces and it looked for a time like we weren't going to need the draft at all.

As for me, I'd wanted to be a writer back as far as I could remember. I signed up for three years in the army, and my plan was I would experience my war like my heroes Hemingway and Mailer had done in times past. You know the idea: The scribbling is always richer and more true to the bone when you write about the things you know first-hand. But right after I signed up, things changed fast. Less than six months later my hero, young President Kennedy, was dead, and for many the blossoming war in Southeast Asia no longer felt like such a grand idea. Personally, I was as distressed as any of us about JFK's tragic end, but unlike many young people, I was feeling more driven than ever to see the war for myself.

At that time I had finished basic training at Fort Ord and was stationed a dozen or so miles south across Monterey Bay at the Defense Language Institute (the DLI), studying in the 47 week intensive Vietnamese course. I was studying hard, but I was also dirt poor and I needed money.

I was the oldest son in my family: my father had died young and horribly of brain cancer, and my schoolteacher mom was somehow keeping my younger brother and three younger sisters in high school and college. So I was sending every dime of my small army paycheck home and looking around for ways to make more money on the side.

Now the army life with its heavily regulated ways seems the last place to go foraging for spare cash; but for me that proved not to be the case. We had a mess hall in our large college dorm-like barracks, and in that mess hall they had a pot or two of terrible coffee brewing all the time. Everybody complained about it, so I thought I saw an opportunity. I went to my squad leader and asked if I could set up a coffee shop in my room. He was a buck sergeant and he hated the army coffee and so he happily agreed to look the other way at our inspections. I went into hock to buy a big 80 cup coffee maker and a carton of the best coffee I could find on sale and hired a classmate who had a car to bring in three dozen donuts a night. Business was so good that I not only could send home more money, for a couple hundred bucks I was able to buy my own car, an ancient faded blue Lincoln.

My first mistake was setting up a second coffee shop in a small cloakroom near the classroom where we had our lessons. I hung up a long rack and everybody from the lowest rank to the highest (we had a Captain in our class) got a free ceramic cup with their name printed on it. The second was not really my mistake, but rather a bit of bad luck when the native head of the Vietnamese language department spotted me unloading a carton of coffee cans from the trunk of my old Lincoln.

"I see you set up shop for coffee in classroom," he said.

"Well, no, it's actually *next to* the classroom. And the army said it's okay."

I felt I was on pretty safe ground here. The truth was, everybody who knew about it was enjoying the coffee and quietly looking the other way. And while the Vietnamese department head nominally ran his department, he was a civilian, and actually not even an American citizen, so his responsibilities were more on the side of hiring other Vietnamese as teachers and handling the departmental paperwork. But I wasn't going to be able to brush him off so easily. He took my arm and suggested we take a little walk.

"I see you have a very fine car," he said. "Your venture must be very profitable."

"It's a very old car," I said. "And I need it to pick up supplies."

"A very fine old car."

"What are you suggesting?"

His disposition soured and he took a step back, frowning at me.

"You should not treat me with scorn. There is old Vietnam saying, *The ant should not kick the elephant.*"

The thought came to me in a flash.

"You want me to give you money! A bribe! A pay-back!"

"I not say that. I NEVER say that!"

I don't know what came over me, but I was really angry as awareness swept over me in a sudden wave of rage. I'm normally a quiet guy and not like that at all, but there it was. I was furious.

"You IMPLIED it! That's why you're sniffing around here, talking about my profits and my fancy car!"

"Khong phai! No, you not right, foolish, silly soldier-boy!"

"I'm just a low ranking soldier, but I'm an American, and you're not! You asshole, you're trying to shake me down! Get away from me!"

There was no more talking after that. He turned and scuttled away like I'd stuck him in the butt with a red hot branding iron. I tried to shrug it off. After all, what could he do? He couldn't boot me out. Nobody flunked Vietnamese class, it wasn't an option. We had some real retards in our class and everybody was expected to pass.

And still, I'd underestimated the little jerk, because in the next few days my 2nd coffee shop was shut down, the Captain explaining he was going to miss his morning cup but he'd been told by his superiors that the DLI had an exclusive contract with an outside vender to supply all the coffee. By then we only had a few more weeks left to our class, so I sold the big pot to an enterprising newcomer and pawned my junk Lincoln off on somebody else and sent the cash home to my mother.

But the elephant hadn't finished with me. At graduation I found out that I had graduated at the absolute bottom of my class, number forty out of a class of forty students.

*In real life my buck sergeant squad leader showed up later in Saigon at the 3rd RRU and appears as a fictional character in some of my novels. He switched over from ASA to the CIA and was very annoyed the one time I took his picture, milling around in the crowd of onlookers the morning after the Brinks BOQ was bombed.

DOM's Place
By John *Mad Jack* Klawitter

February, 1964. The voice sounded over the six foot-high cubicle walls like a foghorn, "Goddamn son-of-a-bitch dink bastards got away again! Last week we send in the B-52's from Guam and we bomb the shit out of them and here they pop right back up again!"

I stopped my translation exercises and tried to listen in. Lowly grunts, E-4 and below, weren't allowed to the Monday morning briefings. Somebody else was explaining something about the Viet Cong being dug too deep underground, and then the gravel pitch voice sounded again, "Well, crap on that, if 500 pound iron bombs don't work, I say nuke the fuckers!"

The meeting broke up and about a dozen officers and civilians made their way past my desk. One, a Buck Sergeant I didn't recognize, made a show of knocking my red covered Hoa's Viet-English Dictionary off my desk.

"Hey! That was no accident!"

Bill Tilgarski looked up from the Playboy magazine hidden in his loose-leaf binder and muttered, "Hey, watch it Sergeant Asshole."

I retrieved my book, looking after the angry stranger, who turned back momentarily to give us the finger and a glare before leaving the department. "What was that all about."

Harry Cone, the civilian head of our section was also leaving the meeting, and he had seen what happened. "That's Sergeant Hinkelby. He failed Ditty Bopper School."

Bill snorted, "Nobody fails Ditty Bopper School. It's Morse Code, for crap's sake. Boy Scouts learn it."

Cone, who was angling for a promotion, tried to be a good guy to the troopers, "Well, Hinkelby froze when those dits and dots started rattling through his brain. He spent three years pushing paper and two weeks ago he re-upped for six more years, hoping to get assigned to crypto."

Bill waved his arms in the air. "That's crazy thinking! How is a guy who can't pop a bop going to unscramble codes?"

"Well, that's what Dirty Old Mary said. She tossed a fit and rejected his application."

I was looking at my Hoa's. I'd retrieved it from the floor with its front and back covers torn completely off.

"A little tape will fix that," Bill said in a voice like that of a mother comforting a child with a broken plastic fire engine.

"But…why is he mad at me?"

"You're his replacement part," Harry said. "Dirty Old Mary asked for you."

I'd seen her around, the messy-looking, grey haired old lady with the hunched-over posture and the voice that sounded like it came out of a horror movie. Mary was a fixture at the Puzzle Palace, a legendary civilian employee who had been a secretary before the second World War, had seen the incredible blunder at Pearl Harbor, had become indispensable (because she knew too much) and so was given the semi-official title of 'crypto teacher' and a job for life. Longevity, in her case, wasn't expected to extend very far into the future because she was a chain smoker who worked in a cloud of grey smoke. She'd have one Lucky Strike dangling from her lips and two or three still smoking in the ashtray at her side.

"Klawitter. Specialist Klawitter. Get the fuck in here." This from around the doorway in her office.

Cone nodded in that direction. "You'll find she's not so bad. Go get her, boy."

After a week, I was chain smoking, too, though I preferred Camels, the brand my father had puffed on right up until his death from cancer.

"All operational crap is dirt simple," Mary was telling me. "Captain Midnight decoder ring stuff."

She shoved the coverts I'd been working on back across the desk at me. In the time it took to glance at each one, she had seen the patterns and somehow found a way to break the codes.

"These guys are low level. Grade school educations." She looked up at me with a warning glance and stifled a cough. "I'm not saying they are stupid, but consider the problems. They need a code they can change every couple days, every week at most. They want to change it before we intercept and decode and nail their asses. What the fuckers don't realize is, we're as interested in triangulating their location as in what the fucking messages actually say."

Her sentence was broken off by a bout of heavy coughing. I was tempted to pat her back, but she held me off with a baleful glare. After a moment and a drink of water from the pitcher at her side, she lit up another Lucky Strike

"Why do you like Lucky Strikes?"

"Reminds me of the war," she said, giving the soft pack in her hand an affectionate squeeze. "*Lucky Strikes Goes To War!* That was the slogan, and they gave them to the men with their rations."

She set down the cigarettes. "Of course, the fucking target was green on the ones the G.I.s got."

"I heard you were here for Pearl Harbor."

"Here?" She gave a little cough and paused as if she might be starting up again. After another sip of water she squinted at me past the line of smoke that was drifting up from her cigarette to join the general haze. "There was no here back then. Just a couple of people in an office in D.C."

She shook her head sadly.

"We saw the signs. We cracked the code. *East Wind Rain.* It was the goddamnedly-ist thing. We were like children. We didn't know what we were doing. Here it is, a Friday and we know the Japanese are going to attack and we have to use a fucking public telephone to contact Hawaii! And it gets worse. One admiral is out horseback riding and a general is having a lawn party and nothing gets through. We finally send a goddam fucking telegram which gets delivered to Pearl Harbor by a guy—a *Nisei,* to boot...there's a goddamn iron clad irony for you—a Jap/American riding a bicycle to warn the ships that hell is on its way from the land of the rising sun, only we are about six hours too late because the bombs are already raining down."

She looked sad and somehow regretful. "If I had been smarter. If I had thought to tell our assholes to send that telegram before we wasted all that time trying to get those fools on the telephone..."

She was staring at a small clear plastic cube with smudgy Ektachrome shots of two small kids on two sides and cats on the other.

"Your kids?"

"Naw. I ain't got no kids. Nieces and nephews. Me, I'm married to the crypto..."

She waved me out of the room as she went into another coughing fit. I was half way back to my desk before she recovered and took in a loud gasp of air.

Friday night Bill Tilgarski talked me into a trip to Dutchman's Mill, a beer hall about twenty miles outside of Boomtown. Tilgarski drove his beat up 53 Plymouth and we showed up during Beer Bash, contests where four or five men sit around a table and gulp down shot glasses of beer from pitchers supplied by the house. The Dutchman's was a big and rowdy hall with round tables and a small patch reserved for dancing. The jukebox was trumpeting *Walk right in/ Sit right down/ Baby let your mind roll on.*

"Come on, Jackie-poo, let's live a little!" Bill pulled me over to a table where the next bash was about to start.

"A couple of Monterey Mary's gonna try to prove their manhood!" a jeering voice hooted, and Sergeant Hinkelby slid into the last chair at the table.

I started to get up, but Bill held me back and said, loud enough to be overheard, "Dumb-ass Sarge is already three sheets to the wind. Don't worry, Jackarino-bear, he'll be the first one down!"

The Bash sounds deceptively easy, but with six men in the contest each agreeing to down a shot of beer ten seconds after the last one, getting plastered fast is a sure thing.

Twenty minutes into the contest, Hinkelby leaned in my direction, "So what was it like at language school, Specialist? I hear the rumors. You have to be a fairy to get in, right?"

Bill, who was sitting in between us, downed his shot and shook his head. "No, you have to be smart. That's why you didn't make the cut." He nudged me, "Ten seconds on you, buddy." I hurried to gulp my beer down."

But Hinkelby wasn[t going to let it go. "Horseshit. Everybody knows you gotta be a homo to go lingie."

"That why you wanted in?" I gave him questioning look.

Hinkelby looked like he was about to leap across the table, but the other four men hurried their shots and it was his turn again. He had to hurry, and he spilled half the glass.

"Re-do!" a passing waitress said.

Hinkelby snarled at her but he refilled his glass and just got it down before the buzzer sounded. He settled back in his chair, seeming to have forgotten the train of thought. In a few seconds he had dozed off, and when his turn came again he was snoring and that was it for him.

Us two *Monterey Marys* nearly made it to the end, but after a while we could not find room for another ounce of Schlitz™, and so we ended up paying for our share of the pitchers. As we staggered out to the parking lot Bobby Bare was wailing on the blaring loudspeakers, ...*Last night I went to sleep in Detroit City/ And I dreamed about those cotton fields and home...* The night air was chill and our breath came steaming out of our mouths.

I noticed Sergeant Hinkelby sleeping in a sparkling new Ford Falcon.

I nudged Bill, and that was a mistake because he yelled, "Hey Hinkelby, you sleeping there all night?"

"You fucking sons-of-bitches!"

Hinkelby started his Falcon, but by then Bill and I had left the parking lot. Bill drove two blocks and decided we both probably needed to relieve themselves. We were waving our dicks in the wind when we heard the Falcon in the distance.

"Oh oh, better zip up fast; here comes, Hinky Dinky," Bill said.

But Hinkelby went roaring by, eyes set on the road ahead.

And yet we hadn't driven five miles before we saw Hinkelby was parked just off the road alongside four or five railroad tracks.

Bill pulled to a stop and I yelled out the window, "Sarge, you can't park there."

Hinkelby woke with a start, but when he saw who it was, he yelled, "Get the fuck out of my life you homo queer lingies!"

That night I dreamed the Puzzle Palace was
a huge barn like place, a mammoth casino and beer
hall with a big neon sign flashing on top that said
DOM's Place. Inside the hall the waitresses all
looked like Dirty Old Mary in every detail except
they had plump young breasts laced up in the front
like the girls at the Dutchman's.

"Dreams mean something," I told Bill on our
way over to the Palace.

"Yes, they do; they mean you are sleeping."

"No. I think Dirty Old Mary is a symbol.
She's quiet, she understands everything, she solves
everything, and she knows everything. I tell you, if
there's one person who stands for this entire
operation..."

"Jackie-poo, you are a certified nutball."

That day Sergeant Hinkelby didn't show up
for work, and about noon the word drifted around
the department he was in the hospital. He showed
up later in the day, banged up and bruised and with
his arm in a sling.

He told Harry Cone the story. "My new
Falcon stalls out on the tracks and I can't get her
started."

"Insurance should pay for that," Harry replied,
the note of false sympathy clear in his voice.

"What insurance?"

"Oh, well, it was defective. The dealership
has to take it back."

"Shit, no," Hinkelby said. "I paid cash." His
angry gaze wandered over to my desk. "I bought it
with the money I got for re-upping for six fucking
years in this man's Army.

*The names Bill Tilgarski, Sergeant Hinkelby and Harry Cone are
fictional. The man I call Bill Tilgarski lost his security clearance, drove trucks
in the bush and after Nam spent years in a VA hospital suffering from PTSD.
Bill appears with still another name in some of my fictional stories in other
publications; I don't remember Hinkelby and Cone's real names, and I don't
think I ever knew the legendary Dirty Old Mary's last name.

My Rise Through the Ranks
By Larry Hertzberg

During the summer after my freshman year of college I hung out, lived at home, and did nothing productive. I spent the evenings with my friends drinking, shooting pool and chasing the other gender. Actually, these were the primary reasons I entered college in the first place!

Sometime in August, a few days before Labor Day of 1963, my dad asked me if I was going back to school. I told him that I didn't know what I was going to do yet.

He said, "I'll tell you what you are not going to do. You're not going to continue to lie around here on your ass. I want you out of the house in two weeks."

Well, after about a week I decided that I had better make a move. I knew my dad meant what he said.

I made a decision. I went down to the Post Office in my home town of Rocky Mount, North Carolina, to enlist in the Coast Guard. Arriving at the door I saw a note indicating that the recruiter had been there the day before and would not be back for two weeks. What now? I didn't have two weeks. I knew I would be homeless in a week so I walked across the hall and enlisted in the army.

On September 11th, 1963, in Raleigh, North Carolina, I raised my right hand and was inducted into the United States Army. The next day I was on a bus heading for Fort Jackson, South Carolina, for basic training. After eight weeks I was on a train from Columbia, South Carolina, to Fort Devens, Massachusetts, home of the Army Security Agency Training Center & School.

I'm a good Southern boy. I had never been north of Richmond, Virginia, the capital of the South. Massachusetts people talked funny and I didn't know the difference in a milk shake and a frappe or the difference in regular coffee and black coffee.

Uncle Sam decided that I was best qualified to serve my country as a 986 MOS. For you ASA'ers who can't remember 1963, that was the Military Occupational Specialty code for a DF Plotter. The MOS eventually evolved and became 05D.

On the first day of class the instructor stressed that the person finishing first in the class usually got his first choice for his duty station. Germany was my first choice and Japan was the second. I studied hard and finished first in the class.

In March of 1964, I got orders to report to the 330th ASA Co., a STRAC unit at Fort Wolters, Texas. I left Fort Devens   where I had been shoveling snow all winter and arrived, a slick sleeve private, at Fort Wolters, Texas, where it was hot as hell.

I found that most of the men in the 330th had either rotated back from an overseas tour, some from Vietnam, and were putting in their time waiting to be discharged, or were lifers. I may have been the only NUG (New Useless Guy) in the outfit.

We did have an interesting mission and I enjoyed working my MOS. In addition to working the mission the Company traveled a lot. We spent the month of May, 1964, in the Arizona desert south of Kingman (Operation Desert Strike) with the snakes, lizards, Gila monsters and scorpions, participating in the largest peacetime joint forces military operation in U. S. history.

As a STRAC unit we were likely to be awakened in the middle of the night, with no notice, to be deployed to somewhere in the boonies for a

few days. Our duffel bags were always packed.
We were deployed to Fort Hood, Texas, several
times for field operations. I got in a lot of TDY. We
had a detachment at Moore Air Force Base near
Mission, Texas, where I was deployed twice for
various lengths of time. I spent some forty days at
Fort Huachuca, Arizona, testing radio equipment.
The best duty I had while in the Army was a six
month TDY to Vint Hill Farm Station, Virginia. Four
of us were assigned to test a prototype radio
direction finding system, the AN/TRD-23. The
operators were William McDowell and Dave
Theuerl. Walt Chisholm was the electronics
technician and I plotted the bearings fixing the
locations of transmitters. We reported directly to
Arlington Hall Station. We were the only DF'ers at
Vint Hill. Almost no one there knew what we were
doing there or what the big round antenna field was
for.

I was on TDY so much that sometimes I went
for months without a payday. In early 1966, I was
still a slick sleeve private. Upon returning to Fort
Wolters from Vint Hill Farm I was informed that I
had been promoted three times while on various
TDY's and was now a Specialist 5. The first rank
insignia I wore on my sleeve was SP-5. I instantly
went from PVT/ E-1 to SP-5/E-5 over night. Wow!
What a payday!

And then there's Vietnam:

In August of 1966 the 330th was flown to San
Francisco where we boarded the troop ship USS
Gen. HUGH J. GAFFEY - Destination, South
Vietnam. For the 18 day crossing I was assigned a
cabin with two NCO's. I had been promoted to SP-
5 just in time for this major benefit.

Apparently, a DF Plotter (now 05D) was a
critical MOS when I arrived in Vietnam in August of
1966. Five days after we arrived at Engineer Hill
near Pleiku, I received TDY orders assigning me to

the 8th RRU at Phu Bai for approximately 30 days. After returning to the 330th at Pleiku, I continued to work my MOS for about three months. During that time I was on TDY to Qui Nhon for a few weeks to set up a new DF site there.

After returning from Qui Nhon, I was assigned new duties. I became a courier of classified material to and from our DF sites and other Radio Research Units around the country. It was almost like being on permanent R&R.

Then one day, to my surprise, I was told that I had been promoted to SSG/E-6. Well, I must have been the only SSG/E-6, 05D40 in the Agency who could not type and could not send or receive Morse code.

Shortly after my promotion to E-6, the Company Commander summoned me to the Orderly Room. I can't remember which one it was because we had several. He informed me that I was now the designated Field First Sergeant of the 330th RRC. My duties? I was in charge of keeping the company area clean and keeping everything in it in working order. I'd report to the Orderly Room most mornings and see Ken Molnar, the company clerk, to claim my work detail. Their most important duties were to police the company area and to apply diesel fuel to the shit cans and stir the shit.

There were a couple major benefits to my becoming the Field First. I got to spend a lot of time at the Company NCO club. I also became the primary operator of a ¼ ton Jeep that was off the books; a midnight requisition. That was a wonderful benefit while it lasted.

One afternoon a NUG Second Lieutenant asked me if he could borrow my Jeep for the evening. He wanted to go into Pleiku for a steam bath and a massage with a happy ending. I told him that Pleiku was off limits but he decided to go anyway. The Lt. said that on his way back to the company area a kid with some type of weapon, or something that looked like a weapon, was approaching him. We think the Lt. pulled the pin on a grenade and accidently dropped it in the back of the Jeep. He immediately unassed the Jeep and it blew up. The next morning he told the CO that the kid threw a grenade and it landed in the Jeep to cover his ass. That's the story anyway.

Just before returning to the world for separation my Great Uncle offered me a $10,000.00 re-enlistment bonus because I had a critical MOS, 05D40. Well, when I arrived in country I had a critical MOS but by the end of my tour DF Plotters had become obsolete. I turned down that great offer and I was honorably discharged on July 27th, 1967, at Fort Lewis, Washington.

So What Brings You to Viet Nam?
By Ron Holt

After I had been seduced into the Army Security
Agency for four years, I was shipped off to Fort
Leonard Wood, Missouri, for Basic Training. That
was September, 1963. During a training session,
one of the Drill Instructors made an announcement.
He told of an event that had just occurred in a
country on the other side of the world; a country that
probably most of us had never heard of. A military
coup had just happened in a place called Viet Nam.
He went on to say there was a good possibility that
a lot of us might go there after AIT. My reaction to
myself of course was, 'Ha, not me. I'm in the ASA.'
After all, I had been told that 'we in ASA' weren't
sent to places like that.

Fast forward: USASATC&S, Fort Devens,
Massachusetts, July, 1964. We had just finished
98C training and were being given our next duty
assignments. Since we had been trained, basically,
on the Russian problem, I had visions of spending
the next year or two with the St. Pauli girl bringing
me steins of beer in a *biergarten* somewhere in
Germany. The Sergeant read down the list of
names and the corresponding duty assignment,
"Anderson, Taiwan; Booth, Okinawa; Cooper,
Thailand; Douglas, Philippines, etc. etc."

Finally, he came to me. I waited with bated
breath. "Holt; Viet Nam."

Viet Nam? Hey, wait a minute. WTF? Out of
my entire class, I was the only one allocated for Viet
Nam. OK, I can live with it. It's not that there was a
choice. Luck of the draw I suppose.

Twelve months later, and it's time to request
my next duty assignment. "Since you've served in a

combat theater, you'll get your choice of assignments" they said.

All right then, St. Pauli girl here I come. All my choices were in Germany. In August, 1965, I reported to Charlie Company, 313th ASA Bn, at Ft. Benning, Georgia.

It was a Sunday afternoon when I walked into the Orderly Room. I had noticed during the cab ride into the Division area that the entire place looked kind of deserted, even for a Sunday. The CQ looked up from his newspaper and asked if he could help me. I handed him a copy of my orders and told him I was assigned to C Company. He nonchalantly replied "They're not here anymore" and went back to his paper.

OK, smart ass, I thought. "Where are they then?" I asked.

"They left Wednesday with the 1st Air Cavalry Division for Viet Nam".

Just Great! The Army sends me from Viet Nam, back to the States, to another unit that has gone to Viet Nam, and I miss them by three days.

I was at Ft. Benning for about forty-five days, and during that time I was generating transfer requests as fast as I could. And, they were coming back denied just as fast. I still had two years left in the army and did not want to spend them in the States. My main request was always Germany, but I was becoming so desperate that even Asmara, Ethiopia, and Shemya, Alaska, was among my requests.

My next stop was Ft. Bragg, North Carolina. And the transfer requests continued to fly and come back just as fast. Finally, at morning formation one fine day, the First Sergeant announced that Headquarters Company, 313th ASA Bn, was allocated to leave for Viet Nam in January, 1966. He also said he realized that some personnel had recently returned from Viet Nam and because the mandatory six months hadn't passed, we could not be sent back involuntarily. But, if anyone wished to waive their six month time and volunteer to go back, that would be just fine with the Army. Shit! OK, you guys win. I'm obviously not going anywhere else.

"What is it Holt?" the first shirt asked when I walked in.

"I want to waive my six months and volunteer to go back to Viet Nam with the company" I replied.

"Wait a minute," he said, "I just saw your name on some orders here". My heart leapt up to my throat. Had one of my requests for Germany been approved? Wait for me St. Pauli girl, I'm on my way!

"Where to Sarge?" I asked.

"The Dominican Republic" he replied.

"The Dom...............?" WTF? Again! After I regained my senses, I thought, watch, the Army will send me to the Dominican Republic, a six month tour, and then, when I come back, they'll send my ass back to Viet Nam involuntarily. So, I waived my time, and waved good-bye to the St. Pauli girl.

And, that's what brought me to the Pearl of the Orient, not once, but twice.

WW III
By Loyd Lavender

WW III. When most folks look at those letters and Roman numerals they think of the war game scenario in Europe played many times with Ivan roaring through the Fulda Gap with more tanks than God. Unfortunately, when I look at those letters I think of one individual. If he thought he could get something out of you, WW III could be as nice as could be, but for the most part he was one of those individuals who made enemies by just meeting him. His enemies would call him 'World War Three' or 'The Turd'.

I first encountered WW III at the ASA School at Fort Devens, Massachusetts. We were both prior service going through Traffic Analyst School in 1964. Fortunately for me, WW III was a class or two behind my class and he was a SP-5 and I was a SP-4. I had been a clerk typist at ASA Pacific in Camp Zama, Japan. I do not know what position WW III was before.

From school, I went directly to Tan Son Nhut, Vietnam. After a couple weeks at Davis Station I was sent up country to Phu Bai, in 1964, and WW III came into Phu Bai a month or two after me. During that time, I had made SP-5. And WW III was not assigned to my shift fortunately. Later on in 1965, two things happened at Phu Bai. We got four brand spanking new second lieutenants to be Watch Officers and WW III was given a new mission that had been assigned from higher headquarters. WW III got both operators and analysts for this mission. WW III immediately became bosom buddies with the second lieutenants. You would not see WW III unless he was in the company of one of the newbie lieutenants. He stuck to them like glue.

To help WW III set it up the new mission, three folks came in TDY, a civilian, an NCO, and a SP-5. After three days the civilian left and went home. The NCO departed a week later. This left the SP-5 who really knew the mission backwards and forwards. WW III didn't like the SP-5 because he knew more than him and because he was a 05H not a 98C MOS. One night WW III got one of his second lieutenant buddies to sign off on a message going straight to Headquarters at Arlington Hall Station saying he was not being supported on this new mission. And being nefarious, WW III did not put a copy of his message in the Read File.

One evening on swings, the Colonel in charge of Phu Bai came in and went to the desk of the new mission and asked a SP-5 friend of mine, who had gotten pulled to the new mission, if he had a certain message with a specific date, time and group number. My friend pulled the message out of WW III's file and handed it to the Colonel. According to my friend it looked like the blood drained from the Colonel's face as he read it. He turned to the Operations Officer and said "don't let that son-of-a-bitch back in Operations."

The grapevine had it that the Colonel had gotten a message direct and 'for his eyes only' from the Commanding General, ASA, demanding to know what the hell was going on. My tour was up a few days later and I left Phu Bai so I do not know what happened to WW III after that.

Two years later, I went back to Vietnam and was processing into 303rd ASA Battalion at Long Binh. I went to the mess hall to eat and there was WW III sitting at a table, still a SP-5, and wearing a patch of an infantry brigade. Needless to say, I was very surprised to see him. I just exchanged pleasantries, since I really did not want to know what he doing, but later found he had latched on to a job at G2 Second Field Force, Lord knows doing what, miles away, I thought I had seen WW III for the last and he was billeted with Battalion at Long Binh. Naturally, when I shipped to the 175th, which was two time, but I was wrong.

I was at ASA Pyongtaek, South Korea, when WW III walked in. This time he was a SSG. I had the misfortune to go to Yongsan, in Southern Seoul, with WW III for a tour of Seoul and Inchon sponsored by the Eighth Army. It was a long couple of days on the tour and was a basic introduction to South Korea. After the tour we were to stay in an Army hotel in Seoul, the name of which I have forgotten. But in the Korean taxi ride to the hotel, I discovered WW III had a deathly fear of Korean taxis. It may have been any Asian taxis, I don't know, but he was deathly afraid of Korean taxis. The funny part of it was, the more he screamed at the driver, the bigger the chances the driver took. At one point WW III was on the floorboard of the taxi curled up and the Korean taxi driver was laughing like crazy. The taxi driver thought WW III was the craziest American GI in South Korea. He may have been right.

The hotel was great though; it had a gourmet restaurant and an excellent bar. Fortunately, that time in Seoul turned out to be my last memories of WW III. I cannot say for certain I was fortunate or unfortunate to have run into him all those times. But I will never forget him terrified, curled up on the floorboard of that taxi. However, I still believe that Saigon traffic was far worse than Seoul.

Chapter 2
Everyday ASA

A Scary Day in Nam
By Juan C. Mendoza

I was fortunate enough to have been stationed in Phu Bai when I got to Nam. However, after a while it got a little boring. You get up in the morning and go to work, you get out of work and head to the trailers then hit the club at night. So when the opportunity came up to volunteer or get volunteered for something different, several of us gladly volunteered and even looked forward to it. One such time was to pull guard duty on the outside perimeter. That sounded exciting.

At the briefing we were instructed to report back to CQ if we saw or heard anything suspicious. We were divided into two men crews, given a radio and driven to the bunker. What I didn't like was the CO saying, "Gentlemen, do not shoot unless they shoot first." *What was that?*

Everything seemed normal and fun; we even got to enjoy dinner packaged the year I was born. Then it started to get dark, and as the sun set it started to get a little scary. There we were, out in the jungle dark as hell with Charlie just a few yards away – for all we knew. And, we had orders not to shoot unless shot at first. What a joke! My buddy and I agreed really fast that if we saw or heard anything out there we were going to shoot first and ask questions later.

It was my turn to be on the lookout. It was just after sunset and getting pretty dark when I saw a light off in the distance. At first I didn't say anything thinking it was my mind playing games on me, but then there it was again. A light turned on then off. This time I got my buddy to look as well. Sure enough after a little while there it was. The light goes on then off. By this time we are pretty nervous. What do we do? Is Charlie coming toward us? Do we call it in or wait a bit and see what happens? We decide to wait. It didn't come on for a while and we were starting to relax a little, but not too much. A light out in the distance at night where there shouldn't be any was something scary and we had to be extra watchful. Then there it was again, it turned on and off. What was odd about it was that it was in the same general location, it had not moved much.

Nonetheless we decided to call it in. The CO, after careful consideration, decided to come out to the bunker and see for himself. After a long time staring out and seeing the light turn on and off we finally realized that the light turned on when a truck passed by on the road to our right and turned off when it was gone. It was the reflection of the truck lights on a can or something. We had a good laugh. The CO wasn't too amused by it and just took off.

The scary part came later when we had the rest of the night to let our minds wander off. Every time we heard a little noise we jumped. We could picture Charlie crawling up to the bunker and attacking us. Just like when you go hunting, if you stare at the mesquite tree long enough you will see antlers moving around it. Needless to say we didn't get much sleep that night.

Forgotten Pisser History
By Mike Little

In 1966, during my first year in Viet Nam, the 337th RR Co. was attached to the First Infantry Division Headquarters near a village named Di An. At that time we were all living in huge squad sized tents and our accommodations were pretty crude. Our latrine was very basic. For a sit down we had wooden outhouses. For number one we had a 55 gallon pee drum with the top off and buried almost flush with the ground. Several small holes were drilled in the bottom of the drum. As the drum filled with urine the diesel oil would float on top of the urine and keep the odor down because diesel oil evaporates more slowly than urine, so you smell little of the nasty stuff, which over time gets ripe in the tropics and can smell pretty rotten. The diesel oil helps but the pisser doesn't smell like roses. This was as good as could be expected. The top of the pee drum was covered with window screen to keep insects and cigarette butts out of the drum, because bugs and butts only make the drum and its contents more foul and odoriferous.

Our pee drum at Di An was out in the middle of our company area central to our tents, near our operations compound, relatively close to our tiny Orderly Room, and about thirty yards from our company club. In our little one-room company club we could buy cold soft drinks and low alcohol beer. Hard liquor was not sold in the club. If anyone wanted to drink liquor they'd have to buy their own bottle at the PX and take it to the club if they wanted a mixed drink.

Our pee drum was out in an open flat area with no privacy screen or barrier of any sort around it. The drum was simply an open drum buried nearly flush with the ground in the middle of our company area. At night anyone who used the pee drum had better be careful as they approached the drum. The street light nearest to the pee drum was perhaps eighty yards away over near the PX. Most guys carried a flashlight at night so they wouldn't trip over a tent rope or bump into something because it was dark as hell around our camp at night.

Everyone in the company was working three rotating shifts, seven days a week. There was no such thing as a day off for anyone. My platoon was the commo platoon and we manned the communications center twenty-four hours a day seven days a week month after month. Everyone was working hard and we needed to relax when we could.

One night the swing shift in the operations compound had just ended and we guys on swings headed for the club to hoist a few *Ba Muoi Ba* beers and unwind.

Our club was tiny; perhaps twelve feet square with one or two tables and a crude bar along one side. We were all talking it up and drinking our beers for nearly an hour. Hofstadter excused himself to go take a leak. No one gave it any thought and we continued to swap yarns and nurse our beers.

Then, after perhaps thirty minutes, someone asked, "Hey, what happened to Hofstadter? He's been at the pisser for an awful long time hasn't he?"

We suddenly realized Mr. Observant was right. So we organized a search party of drunks to go find Hofstadter. A couple guys had flashlights so they led the drunken posse out to the pee drum. At the pisser we were alarmed when we found the screen that covered the top of the drum caved in and torn away. The ground around the drum was soaked with what had to be urine and diesel oil.

45

When we saw the screen and the wet footprints leading away from the drum it didn't take a forensic scientist to conclude Hofstadter or someone had taken a dive into the pee drum.

We followed the soggy footprints to a tent. There we found Hofstadter unconscious on his cot wearing his urine and oil soaked clothes. The mystery was solved. As disgusting as this sounds, we decided to leave him be. If Hofstadter was comfortable sleeping in piss and oil soaked clothes, why should we disturb him?

The posse disbanded and most of us returned to the club for a final beer and ruminating about Hofstadter's midnight swim. Then it was off to bed for everyone. Mystery solved.

The following day some enterprising soul dug the post holes and used 4 x 4 posts to support a canvas privacy screen around the pisser so no one would take a midnight plunge again.

After the privacy screen was up someone ran a power cable from the motor pool next door. The cable was used to power a single light bulb that was hung above the pisser.

With the pisser screened and lighted someone hung a crude cardboard sign below the bulb that said, 'No swimming while lifeguard not on duty.'

This happened more than forty years ago but I still smile when I think of that damned sign and Hofstadter's midnight swim. We gave Hofstadter grief about his plunge for months but he just smiled. He finally rotated back to the States, and that was the end of it.

Midnight Run
By Mac McDaniel

In 1967, I was assigned to the 330th RR Company on Engineer Hill outside of Pleiku, South Viet Nam.

We would transport boxes of raw collection, three pages of the five ply paper with raw intercept as well as tapes and other courier documents, to the Pleiku AFB via 'Midnight Runs.'

On one famous memorable night, SSG H was in charge of the convoy and left Engineer Hill with the Jeep in the lead followed by a 3/4 ton fully loaded with boxes. The normal procedure was to have a Jeep behind the 3/4 ton to ensure nothing fell out of the back of the 3/4 ton. They arrived at the courier aircraft and unloaded the boxes. However, there was a major problem. Half the boxes were missing.

Back at the 330th RR Company, a large search party was assembled to search for the missing boxes full of classified information. It was a little after midnight so odds were in favor of finding the missing boxes. They searched the road and sides of the road all of the way from the 330th RR Company area on Engineer Hill to the Air Base where the courier aircraft had been parked. No boxes were found.

A couple of weeks later, an ASA Special Forces soldier from one of the SOD units, a guy familiar with our raw intercept, was at an ARVN camp and noticed the ARVNs had supplied their outhouses with used three ply paper.

The mystery of the missing boxes of classified intercept was solved!

The SSG H who was in charge of the 'Midnight-run' that lost the intercept was dispatched to the ARVN camp and had to inventory all pages of

intercept included those that had been used twice if
you know what I mean.

Snakes in the Antenna Vaults and Ducts
By Steve Polesnak

We all realize that working in an Operations Building in Southeast Asia can be quite an experience. Throughout three tours in Asia I became aware that the snakes of Asia were quite dangerous and venomous. Cobras, banded kraits, and vipers were very venomous and everywhere. We found them in the tents crawling up through the pallets, in the raceways for the antenna cabling, around the entire perimeter of the operational areas, and on the streets at night when we walked to work. We also saw many insects that could do a job on the body with their bites. You had to be on guard at all times.

Operators on their positions always had to worry about the little slithering varmints coming up through the ducting/raceways because the antenna wires from the outside of the building ran under the floor of each operational area. It wasn't unusual to find baby cobras in the ductwork and most operators kept a stick or pole to kill these very venomous reptiles, before the operator was bitten. To say the least, operators kept their eyes and ears open while at work in the intercept bays.

All during my tours in Southeast Asia I had always wondered why I had never heard of any of our maintenance people being bitten by either the larger or smaller snakes in the vaults which were built from the operation building, to the antenna. The vaults had to be checked quite often to ensure there was no deterioration of the cabling, or water in the ducts. So for many years, I just wondered how these maintenance actions were performed.

Forty years after my leaving Viet Nam and Thailand I asked one of the people who was the supervisor for following up on all maintenance issues and keeping the vaults clean and the antennas working, why I never heard of him or any of his people getting bitten while performing their maintenance activities.

He said to me, "I will tell you our secret."

"First of all," the supervisor said, "I would tell you that I wouldn't get my ass or my maintainers in those vaults anytime due to the number of cobras and banded kraits who could get into the cabling vaults and the Operations Building. When it was time to do our maintenance tasks, we had a couple of specially trained local nationals under contract full time with the responsibility for them to go into those vaults, and specifically check and clean out any non-human entities which might hurt our maintenance persons. Do you know how much it costs to train those Army maintenance people?"

When maintenance was to be performed, the local nationals started from the antenna proper, walked or crawled through the vaults all the way back to the Operations Building, and then when ready, they would call me and I would release the maintenance personnel to go to perform their functions while the cabling vaults were clear.

He ended his explanation with, "My mother didn't raise a dummy!"

Military Intelligence, Oxymoron
By John Stone

When I got to FORSCOM working with G2, my wife called me one day and said I know now what you are. I asked her what she meant and she said that on the talk radio that day someone had asked for callers to call in their oxymoron. The first caller said it was Military Intelligence. Not a bad call.

During my 12 month tour of duty in Vietnam, I served with the Army Security Agency (ASA) at Pleiku. The ASA was a unit that officially was never in Vietnam until very near the end of the war. That's why you may hear ASA vets say, "We were never there."

At Pleiku we were working 12 hour shifts on and 12 hours off. As Sgt.-in-charge, I had given some of the linguists a break to visit the club and get something to eat. They had been gone about twenty minutes when one of the 05H's had something he thought was important and needed a linguist to look at. Of course, they were all at the club. I sent a man to get them and in a few minutes they all came back, looking kind of scared and one of them said, "Sarge, any time you feel you need to call me back, please do so."

I asked what happened and they said as they were leaving the club a white phosphorous mortar round fell short and hit the table where they had been sitting. Just lucky, I guess.

As a Sergeant E-6 assigned to ASA, I was required to participate in activities outside of the Pleiku base. During one of these times, I was at the base at An Khe for almost a month. During a short break one of the Specialist E-4's working with the three of us from Pleiku, asked me if I wanted to go to the exchange. An exchange is the military term for about the only store that was safe to shop in because it is run by the Army Air Force Exchange Service (AAFES).

After I told him yes, my intelligence should have kicked in when the Specialist said, "We will go first thing tomorrow." But that was my military Intelligence and sometimes it is slower than real intelligence.

As the Specialist and I got in the small truck for the trip to the exchange, we started out the main gate, I asked, "Where are you going."

It was only at that time he told me, "The Air Force has the best exchange at their base so I thought we would go there." Again, my intelligence should have kicked in and I told him the Army exchange would do, but again, my military Intelligence.

About an hour from base the truck started not running as well as it should and I could just see the horrors of myself and one man, with two small guns, trying to get back to a safe place. He said there was a small DF base near us, called one rock ranch, and we would stop there. These small DF bases were manned by usually less than ten people with a Sergeant E-6 in charge. As we drove into the base there was a small helicopter sitting in one corner of the base.

We went onto the base and while the Specialist was fixing the truck, I was talking to the Sergeant-In-Charge. We talked army talk for a while; then he asked me, "Sarge, is there any special activity going on that I need to know about?"

I told him no. Then I asked him, "Why?"

He said that for about a week to ten days they had been hit by sappers almost every night. Sappers were VC soldiers who would attempt to enter your compound at night with explosives and blow up something.

It was about this time that I asked the million dollar question, "Where did that helicopter come from?"

Laughing, the Sergeant told me, "Boy, did I make a good trade. I traded one of the air conditioners out of one of the expandable vans to a base down the road for the chopper. I am learning to fly it, he said. I can only go up and down right now but I hope to learn enough to get my license when I get back to the States."

"How long have you had it," I asked him.

"Oh, just a little over two weeks now. I think I am doing good just to get it up about 100 feet off the ground in this short a time."

My military Intelligence finally kicked in and I told the Specialist, "Get this truck fixed; we need to leave this place."

He fixed the problem and we went on to the Air Force Exchange. We spent less than an hour there and I never was so glad to get back on the base at An Khe.

I never did learn if the Sergeant ever figured out the connection to his helicopter flight training and his nightly sapper attacks, but for me, it did not take military Intelligence to put two and two together and to get the hell out of there.

Our special mission at An Khe took us to work on some of the Fire Support Bases along the main road running out of An Khe. These two bases, one called Action and the other English, kept us working to help find VC patrols or units by listening for their radio transmissions and then have them located through DF and other methods. One of my greatest war moments came when we were traveling from LZ English to LZ Action during this mission.

An Officer assigned to the Calvary unit stationed at An Khe had given us a command track along with a PFC driver from his unit. A command track normally carried about 10 to 12 personnel with a .50 caliber machine gun on top. We had traded the 10 - 12 people capacity for a few radios we used for our mission. While we outranked the PFC and we had been in country longer, this PFC was probably the most knowledgeable soldier of the area and only one of us who had seen any combat.

On this return trip we met a fairly large convoy, with one vehicle broken down on a narrow bridge. The PFC looked to me, the highest ranking guy, playing John Wayne on top of the track holding on to the .50 caliber like I was going to win the war by myself, and asked, "What do you want me to do?"

Dumb, I thought, "Go around," I said.

He replied, "But Sarge, it might be mined."

"Don't look mined," I said, and so he left the road to go around the bridge and we continued on down the road.

That was one of the dumbest things I said during the war. Who could look at the ground and know it was not mined? As it turned out, it was the best thing I could have said. About two or three miles down the road we saw a lot of helicopter gunships going overhead. When we got back to LZ Action we were told that a large VC unit hit the convoy we had passed at the bridge. Were we just lucky or did my military Intelligence save us? Perhaps military Intelligence is not an oxymoron when it saves your ass.

There Ain't No Such in 'Nam
By Robert Flanagan

In the fall of 1964, I lived in an NCO hooch adjacent to the Orderly Room at Davis Station on Tan Son Nhut Airbase, northwest of Saigon. The 68th Armed Helicopter Company, who'd only recently received the new UH-1 'Huey' helicopter, was our next door neighbor. Their ramp—separated from the 3rd RRU by a corrugated metal fence which did nothing to shield us from a 'round-the-clock subjection to toxic levels of rotor whine and missing engine backfires—was a beehive compared to the lassitude of ASA's benign presence.

   One evening, lying on sweat-soaked sheets under my mosquito netting, reading, I was drawn from the hooch by the sounds of shouting, stamping feet, slamming screen doors. I walked out in my skivvies and saw SP-5 Roger Cook, an analyst from White Birch, soundly but verbally assaulting the CQ. "Get me a Jeep," he demanded. "I gotta get to the hospital."

   "Whassa problem," the CQ, a disinterested SP-5, asked laconically.

   "I got stung by a scorpion!" Cook said. "Hurry."

   "I can't get you a vehicle just for that. 'Sides, there ain't no scorps in Viet Nam."

   "The hell you say. Get me a ride, man. Now!" Both men were SP-5s, but it soon became obvious that Cook would prevail. For one thing, he was bigger, and—as he shuffled about before the CQ's desk, standing on first one foot, then the other, all the while tightly gripping his right hand in his left—he was angrier.

   After a few false starts, the CQ acknowledged he'd called the motor pool and a vehicle was on its way. Cook stormed outside. I thought he might need help and stepped back in my hooch, drew on some cut off fatigues, and was just behind him when he jumped into a three-quarter ton truck. I told the driver to get us to the 68th's dispensary, a small flight medicine shop that was nearby and manned around the clock. The driver, who'd assumed an attitude of solicitous help watching Cook in obvious pain, got us there in less than two minutes.

A Spec-4 in greasy fatigues manned a footlocker desk inside the shabby hut. There was no room for me inside, so Cook went in alone and I sat with the driver for a couple of minutes, then realized that as we'd gotten Roger to medical aid, I served no purpose, got out and walked back to my hooch. The repetitious dialogues that followed were related to me later by Cook during another mutual adventure. (More about that another time.)

Cook walked up to the Specialist and said, "I need to see a medic."

"You're looking at one." Not convincing, but given the venue...

"I got stung by a scorpion."

"Sshhh— ain't any such in 'Nam, man, the unlikely medic announced.

"Don't 'man' me, man. A scorpion got me on the hand. I need some help." He looked about, seeking a more substantive medical opinion.

"Well, don't make any difference. I got nothin' here to help you with bug bites 'n' all. A door gunner catches a round through the lungs, I can put a pressure bandage on, do a blood test, wash wax outta your ears. Cobra bites and—"he snickered nastily "—scorpion stings' not in my job description."

The driver, inside by then wondering at the hold-up, asked, "Where can I take him for treatment?"

"Try the Air Force 32nd Dispensary. Not far, right here on Tan Son Nhut."

"Yeah, I know it. Thanks ever so much," the driver added snidely. "Come on Rog; I'll get you there. How're you doin?"

"My right arm, now ... in addition to the bite site on the hand, my whole arm is burning. And swelling."

The driver looked down at Cook's rapidly inflating limb. "Let's go!"

Back in the three-quarter, they sped across the unlighted spaces on the airbase and screeched to a halt before the Air Force dispensary. The bright glow of lights, within and without, promised better than the helicopter Band-Aid shop. Inside a tile-floored lobby, at a neat desk, the neat Airman first class manning the desk asked if he could help.

When Roger explained, the Zoomie looked skeptical but before he could respond, a Major in something less than elegant officer's dress spoke up: "Soldier, there are no scorpions present in Vietnam. Those kinds of threats are the province of hot, dry desert environments. Leeches, snakes, many other invasive vermin, are here but not scorpions. Come back, let me take a look and see if we can determine what has accosted you."

"Sir," Cook insisted, this was a scorpion. I know from experience."

By then, the doctor had peeled back Rog's sleeve and seemed surprised at the distended flesh and angry red streaks up the soldier's arm. "Well," he said, "something has certainly taken a dislike to you. Could it have been a small viper? Did you see anything?"

"Shit ... sir, I saw a goddamned scorpion. I been there before." At this outburst, he reached into his vest pocket and pulled out a penny match box. Holding it awkwardly in his swollen hand, he pushed from the far end, the box slid open, and

Roger tipped out onto an examining table a visibly angry and hyperactive scorpion, the tell-tale tail erect and arched over its back, poised to strike. "This scorpion's the one I saw."

"Hell, that's a scorpion!" the doctor said, astonished. "That can't be right." But the evidence staring him in the face was irrefutable. "How'd it happen?"

"I was rolling down my mosquito net, getting ready for the sack, when I felt it strike. Musta been in the roll of netting. I knew the instant it struck what had happened. I got stung at Fort Irwin, California, three years ago on training maneuvers. The same poker-hot surge of pain, immediately." Cook didn't embellish the experience.

The Airman first said, "Why'd you put him in the matchbox?"

"Just because of this attitude I got here, from the CQ, the medic at the 68th, and now you guys. Doesn't anybody in this frigging country know about these things?" Through his elevated sense of abandonment, it was obvious Cook was feeling a constant increase in pain. He said he could watch the arm expand in girth as the babble went on.

In the background, the doctor had gone swiftly to a bookshelf and removed a medical book, thumbing rapidly through index and glossary and reading from relevant pages. "Is Airman Phelps on duty?" he called out.

"Yes sir, in the lab I think, he's catching up on some water purification analysis."

"Get him." He explained lamely to the small crowd, "Phelps knows a bit about tropical critters. Maybe he can advise something. In the meantime ..." he indicated the reference books.

When the critter-gitter Airman Second Phelps came and was asked, he replied, "Of course, there are dozens, maybe hundreds of varieties of scorpions throughout Southeast Asia. There's an almost cult-like behavior here in Vietnam in eating scorpions, and they're bred, sold and traded for that market. But there are many species in the forests, the open country, and the jungles. It's not true they're only found in desert country. Now, this little fellow is pretty small, and with small pincers, but strangely, he's likely one of the more venomous. Some of the species *Heterometrus laoticus* grow to seven-to-eight inches in length. Many are bred for fighting, and scorpion combats commonly feed the Vietnamese penchant for gambling." Suddenly losing interest, likely having expanded his range of scorpion knowledge, Phelps wandered off.

Though the Zoomie doctor eventually discovered a prescriptive treatment, and that he had on hand the necessary drugs, and gave Cook a series of injections which, over the course of several more days, eventually reduced the swelling and pain, and reduced the threat of death by the scorpion-that-wasn't there, no one in the 3rd RRU who heard about Rog's travail thought much about the Air Force's tropical medicine program.

The Trinity
By Robert Flanagan

Back from a clandestine stroll through the bush with the 372nd and a LRRP team from 3rd Brigade, we adjust quickly to big, bold Camp Evans, now base camp for the Screaming Eagles after a history of tenancy by the Marines and the First Air Cav. Halfway between Quang Tri City and Hue, we're content to sit here beside Route 601 and travel neither to Phong Dien nor the shabby insubstantiality of LZ Sally. We feel especially blessed in his holiday season, though: better here than out in those recent cold misted mountains of Eye Corps. Semi-regular mail, hot showers, and chow you don't have to coax with a plastic spoon out of a can—it is unreal to the point of epiphany-a sort of ephemeral kind, to be sure. This too shall fade away; but for now...

Hit the showers fully clothed, or as much as those tattered rags permit the description, in my outdated no-see-me suit, with web gear and weapons. Strip off the tiger-striped, rotted fatigues and jungle boots, throw them straight into the trash can with a touch of wan regret. I'd traded two cartons of Salem's™ to a Zip Marine for those threads in order to be Mister Cool out beyond the wire; and it wasn't easy finding my size in the South Vietnamese logistical chain.

Following the lengthy, almost-sexual enjoyment of the shower, I hop-scotch, naked, back up the 110-degree F. cinder street to the temp BOQ hooch. It has been so long since I've had my boots off, the cinders do a job on my bare feet. But it is a two-edged sword: the unconscious act of bravado rids my feet of layers of sloughing, milky-gray skin. In IV Corps, my peds would be worse from immersion foot. Trench foot, my grandfather of WWI familiarity would have called it. But we have no trenches.

Quick patch through the M-16's bore, light coat of oil on the barrel and receiver assembly; spread out the web gear on my bunk to dry, and head for the chow hall. Passing Sergeant Major's hooch I hear Christmas carols—the Harry Simeone Chorale. Probably a tape from home I think. The schmaltz diverts me to the mail shack.

Four letters, a double-shoebox-size carton, and a flat pack I knew would be newspapers. The letters are Christmas cards from *der Frau und Kinder*, scholarship evident in Sean, the five-year-old. Picasso from Liam, age three. Passing, I catch a whiff of swill from the direction of the mess hall, realize I am famished, and hold off on the packages. A good thing, that; I cannot handle ecstasy on an empty stomach.

After the less-than-exhilaration of bad Brunswick Stew on overcooked noodles, all the while facing Monroe across the table with his thousand-yard stare, gone Asiatic after receiving in today's same mail call a Dear Clyde letter from Murtha down in Nacogdoches, I return to the hooch and the packages with anticipation. Thinking—See, even with all his time-in-grade, a senior Captain, Monroe wasn't secure from the snickers of fate, the fickle finger thereof.

I'm excited, thinking of Christmas, three days off, and wondering what the home front has provided by mail that will be useful/ welcome/ interesting/ cheap, in this place where normal emotions are often reduced to indigestion. Like any child, I tear into the carton first. Wow! *Shazam!* My tired, dust-strained eyes cannot assimilate such wonders to behold. The box is filled with a mélange of purple paper packets, small tubes of cardboard-wrapped something's, and a couple of white, opaque plastic containers. I immediately scan the area, ensuring a state of isolation. If the news of this windfall gets out, every trooper—FNG to jaded vet—will be on me like stink on a rice paddy.

The Trinity! In one swell fop. Answers to prayers of even limited issue. Onesy-twosy-threesy? No, this miracle is here, *completo. Magnificato!*

Grape Kool-Aid™

McIlhenney's Tabasco sauce™.

Ammens Medicated Powder™.

In this world that I find myself in—where every microbe is another enemy; where the NaDCC has replaced halazone tablets and will counter those enemies, but still flavors the water with iodine—drink the iodine, even take the water intravenously. But no, those innocuous little nickel packets of grape Kool-Aid will stifle the gag reflex enough to keep my body hydrated. The grape does nothing to disguise the sludge nature of the water, often filched from the ochre baths of water buffalo and turgid streams where every kid in Zipland, as well as the boo make water; but consistency aside; it covers the iodine which covers the slime. The carton must contain a hundred packets of that wonder powder, serving as packing for the other treasures.

The cardboard tubes can only be one thing, the size being so fondly remembered: Louisiana hot sauce! Cajun syrup. And the best of the lot, I find, as I rip the cardboard from one of the small tubes: McIlhenney's. From New Iberia. Just having driven past those pepper fields, I didn't suffer from sinus trouble for years after. McIlhenney's, to cover, disguise, eliminate, change, rid me of the tastes of the foulest indiscretions in the mess cook's repertoire. First the Grape elixir; now this. My head swims with the richness of my blessings.

And then, the Holy Grail; lying cool, white, and benign, no odor evident, two plastic containers of Ammens Medicated Powder glow with alchemist pride. Here is the answer to the ills of the Orient; the end to body rash engendered by wading neck-deep in God-knows-what cornucopia of filth and disease, paddies of human excrement. The end to prickly heat caused by the prickly heat—the fetid breath of this miserable climate. But above all, the answer to the romantically labeled *crotch rot.* The inevitable, ubiquitous sloughing away of flesh in the moist areas of the body, comprising underarms, beltlines, boot tops, and other bodily creases, but overwhelmingly aggressive in the region for which named—one's crotch. Trapped moisture, given no chance to air, never to dry with the eternal demands of foot-slogging through the downstream wet of a monsoon country, causes skin to wrinkle, to swell, to fester, to blister and, following the first breaking of the cysts, in that fecund atmosphere the unbridled growth of every form of laboratory subject: germ, virus, bacteria, algae, fungus, even tiny mushroom growths, which rip the skin from the infected areas faster than napalm burns but have much the same feel.

All of us, whether on the ground or flying, face this common enemy due to the nature of the uniforms our military service dictates. Wherever sweat arises and is sustained, and air is cut off, the malady appears. Charlie doesn't get crotch rot . . . but then, he barely has uniforms. I'd tried baby powder, Mennen's, rice flour, Lifebuoy and Octagon soap, homemade lye soap, and from the dispensary dozens of unguents, salves, topical dressings, penicillin wipes, injections and ingestions, and a whole bewildering complex of compounded drugs and applications. Nothing worked for me. Sometimes, one sufferer or another will stumble upon some mix of lab ingredients that work to some degree, for him. Word gets out. Everybody tries that mix. It never works for anyone else. But for me—nothing.

That is until Red Barbizon saw me cringing in the shower one evening after a particularly long and exhausting flight. I was in tears as I forced myself to wash those tender, favored nether regions. "You pick up a dose?" he sneers.

"You know better, you vicious asshole. Grade A crotch rot," I snarl back.

"Man, young Sergeant, you silly or something? You ain't using Ammens?" He stares at me as if one of us is on leave from a mental health ward.

"Tried it all, Red. Mennen's, baby powder . . ." I name off the failures of my discomfort and continue washing through tears.

To cut to the chase, I had somehow never heard of the efficacy of Ammens for this particularly Job-like visitation. I race to the PX where, miraculously, they have it in stock; I cover my body so that I look like one of those outback Aborigines, whitewashed for fertility rites. And within days, the skin is healed, the pain, itch and swelling gone. From that day 'til the end of days, my medicine cabinet and travel kit will never be without Ammens.

I have the three items that will sustain me through the nine remaining months I have in 'Nam at that time: Grape Kool-Aid, McIlhenney's Tabasco, and Ammens. And despite a soul that is belligerent and caustic on the subject of conventional religion, these near-miraculous icons of tested belief stand at that supreme elevation: the Trinity!

The Shadow Bar
By Lee Bishop

Coming into Pleiku from the 330th Radio Research Station on Engineers Hill outside the city in the Central Highlands, I usually stopped by the Shadow Bar for a drink or two and a chat with the bar girls, especially to talk with the beautiful Li-Li.

Li-Li was small of stature even for a Vietnamese. She had skin that was rice powder white and appeared nearly translucent. She was a sloe-eyed beauty who could melt any man's heart. Her dark, soft hair fell to the middle of her back. Her lips were almost too perfect to be real. I thought she was the essence of femininity, and every man I ever saw look at her did a double-take.

Jumping off the back of the deuce-and-half truck, brushing through the crowd of peanut girls, I headed to the bar. It was a beautiful day with sunlight dappling the shaded roadway.

"I wouldn't let you smell my pussy for five P!"

That caught my attention. I turned to look back at the pre-teen girls selling small paper cones of peanuts wrapped in brightly colored tissue paper. One of the GIs jumping on the back of a truck for the trip back to camp had apparently offered her five piasters for a package of her peanuts. Obviously she was offended. And just as obviously, she had saved face as the soldier's pals were laughing and slapping him with their hats. The peanut girl was a heroine among her friends that morning.

I resumed my mission and approached the Shadow Bar. Shadow Bar is an unusual name that had nothing to do with the building itself. The walls were bright, white stucco with the front of the one-story building evenly divided by a massive black door with a small glass window centered about five feet up from the sidewalk. The few windows in the building were covered with shades and draperies that allowed no glimpse of what was happening inside.

I entered and as my eyes adjusted to the dim light. I dodged a couple of tables, made it to the bar, and asked the bartender for a Ba-Muoi-Ba. Among GIs "33" was the most popular of the Vietnamese beers. Perhaps Ba-Muoi-Ba's greatest claim to fame was its formaldehyde content. Occasionally someone would get a "green" 33, take a sip, groan and sputter, and spit it out. "Green" meant that there was an error on the production line and someone got more formaldehyde than beer.

"Anh Oi, lay day nhe? ("Come to me, older brother, will you?")

The girl in the dimly lit corner of the sitting room motioned me over to her; "You are in danger, Anh Lee. You must be very careful, and you must not get too close to Li-Li."

"I never have, Phuong, but I have to admit that she is very beautiful, and she seems to like me. What is wrong?"

"It is because she is so beautiful that the VC have made her sick. If you stay with her then you will have the disease also."

"Are you sure?" I asked.

"Soon she will ask you to spend time with her. Then she will introduce you to the owner. He is VC. He will ask you many questions. You must not tell him anything, but you also must not let him know that you suspect him. If you do, you will die. I must go."

This seemed weird, but everything about my journey from a small town in northern Ohio to a city in the Central Highlands of the Republic of South Vietnam was unexpected for a kid who had graduated in a class of thirty-three students in 1963 less than three years earlier.

How the hell did I get here in the first place? After a year at Ohio University and a summer working at a Ford truck assembly plant in Wayne, Michigan, I returned to my home in South Amherst and took a job with the Nordson Company as a tool & die maker's apprentice.

A couple of guys that had graduated from high school with me picked me up at the end of my afternoon shift, and we went drinking in Lorain. We ended up in Hannah's Bar on North Broadway just a couple of blocks south of Lake Erie.

Hannah's was just a hole-in-the-wall, long and narrow with a bar running almost the whole length of the building. The brick on the face of the building, and the white wooden door outside were nearly black from years of soot from the coke plant and steel mills. Patrons tended to be older folks who spent what was left of their paychecks on booze, Polish sausages, hard-boiled eggs, and illegal tip boards.

Truly the only thing Hannah's was good for was drinking, and despite our ages we ordered up and were served the working man's traditional shot-and-a-beer. In fact we were served a lot of tradition that night, and by two in the morning we were beginning to feel our oats.

Curtis was so happy he pulled his gun and sent a round down the length of the bar and into the back wall. Hannah was upset but didn't dare call the cops since we were obviously underage. He sent us packing and told us to never return. This would be no problem, for Curtis and Stanley had a plan.

During the course of the evening, they told me they had enlisted in the Army. Their new plan was pretty simple. They had decided that I should go to basic training with them on the buddy system.

Well, since that was the only plan the three of us were able to come up with that whole evening, I figured it was my destiny. There really wasn't anything else going on in my life so as dawn arrived we headed for the Army recruiting office. There were, of course, some tests for me to take.

Apparently the night with no sleep and lots of alcohol didn't hurt things too much, because Sgt. Grace came back all smiles, shook my hand, and said; "Son, we're going to make you the James Bond of the United States Army!"

"Two things, though, I need to tell you about, Lee. We will put so much money into your security clearance (six figures) and your training (ended up being a year in language school in Monterey, California), that we'll need you to sign up for four years instead of three."

"Anything else I need to know, Sgt. Grace?"

"You won't be able to go through basic training on the buddy system with your friends. They'll be going to Ft. Knox. We send all Army Security Agency personnel to Ft. Dix, New Jersey."

"Tell me more about this spying business, Sergeant."

"Well, now, that's nothing to worry about. You'll be a key and critical component in your country's defense system, but even if you go to war you'll be so far behind the lines you'll never even hear a shot fired in anger."

I need to digress a bit and tell you something that you already suspect: Sgt. Grace lied. First, the language I was assigned to study was Vietnamese. Second, after going through the 101st Airborne Division's jungle combat school in Phan Rang, I was assigned to the 1st Brigade, a reactionary unit. I joined them in Dak To, and early the next morning was flown out to join an artillery battery in what was called "Operation Eagle Bait." It didn't take long to find out we were the bait, and Charlie was the eagle. The objective was to tempt the Viet Cong into attacking us, and then bring in an assault wave of Huey's loaded with infantry and wipe the enemy out.

What I quickly discovered is that there is nothing in the world comparable to the first night you are brought out of a deep sleep by M-60 machine gun fire.

Anyhow, I took the sergeant up on the deal he offered. After three months with the 101st in both Dak To and Tuy Hoa, I was sent on a temporary duty assignment (TDY) near the DMZ to learn field station operations from the 8th Radio Research Unit at Trai Bac Station in Phu Bai. From there it was off to Pleiku where we worked first in support of the 25th Infantry Division and later the newly arrived 4th Infantry Division. But let's get back to the Shadow Bar and the lovely, if perhaps diseased, Li-Li

"Anh Lee! Lai day. Li-Li yeo anh nhieu qua." ("Elder brother Lee. Come here. I love you too much.")

Having just spoken to Phuong about the danger I was in, seeing Li-Li held less joy than I had been anticipating when I jumped on that truck earlier this morning back at the 330th.

"Li-Li! I was looking for you. Let's have something to eat."

"I have to work, GI. My boss thinks I spend too much time with you. He would like you if he knew you. Yes! Come with me. I want you to meet the man who owns the Shadow Bar. He is very important. You will like him."

Li-Li led me into a corridor leading to an office in the back of the building. There she introduced me to Mr. Nguyen van Duong who wasted no time in getting to his questions. Duong was an ugly man, small and fat at the same time. His hair was thin for a guy in his 40s, and he used a pomade to plaster it to this head. The skin on his face had an oily sheen to it. His lips were fat and reminded me of raw liver.

Duong dismissed Li-Li and started asking me about my job.

"I'm just a clerk."

"How many Americans are in your unit?"

"I don't know; quite a few."

"Why don't you know more about these things?"

"I'm just a clerk."

The questions continued and I kept dodging them. Finally he dismissed me with a frustrated wave of his hand. I went back through the corridor that Li-Li had led me earlier, but I didn't see her when I returned to the bar. In fact, I never saw her again nor did I ever return to the Shadow Bar.

These odd moments at the Shadow Bar happened more than forty years ago, but I still wonder if I should have reported the incident when I returned to Engineers Hill.

I chose not to for a couple of reasons. Most significantly, the whole experience seemed oddly surreal. If I could hardly believe it, who else would? What's more, it did not strike me as an incident worth losing my security clearance over. There was no reason for my clearance to be lost, but stranger things had happened. I thought I should be cautious.

Besides, I learned in basic training that volunteering information never went unpunished.

You Have Venereal Disease
By George Murphy

I was at the 3rd RRU from April, 1964, to April, 1965. In early 1965, I noticed a lump on the right side of my groin. After ignoring it for a number of days, I went on sick call. I explained to the medic my symptoms. The medic told me that I had a venereal disease and he could give me some pills that would not go on my record. I replied I know how you get venereal disease, and I do not have venereal disease. I want to see the doctor.

After a wait, I saw the doctor and he said I had a hernia. He was going to send me up to the Army Field Hospital in Nha Trang for a hernia repair. The doctor handed me my file and said schedule a flight with the medic. I walked out and handed my file to the medic and I said 'hernia' and I need a ride to the Field Hospital.

Several days later I was on a flight from Tan Son Nhut to Nha Trang. I hopped a ride to the hospital. As I remember, the hospital was a tent with a wooden frame. That day I saw a surgeon and he confirmed the hernia and I was admitted to the hospital. A medic told me they would work me into the mix.

After several days, the doctor came to see me. He said that there was some action up North and they were expecting wounded and they would need my bed. They were sending me back to my unit. I asked what I should do. He said you should be able to push the hernia back in. If it becomes strangulated, it will be an emergency and we can fix it. Later that day, I was on a flight back to Tan Son Nhut.

Several weeks later, I was back at sick call. I think it was for shots. The doctor, who sent me to Field Hospital, asked me how it felt having the hernia repaired. I explained what happened. He said that they are not supposed to do that. Then he said I am going to send you to the Evacuation Hospital in the Philippines.

The next morning I walked over to sick call and I was driven to the airstrip. There was a C-123 parked on the apron. It was surrounded by ambulances. Medics were loading the wounded into the aircraft. Everyone was on a stretcher with an IV. They were stacked two high on the aircraft. Other than the flight crew, I was one of two people who could walk onto the plane. The other guy had some kind of flesh eating bacteria. Needless to say, I moved to the web seat on the other side of the plane.

We landed at Clark Air Force Base. It was like a scene out of MASH. The wounded were carried off the plane and the medical staff surrounded them. They were loaded into the ambulances. Everyone was admitted to the hospital except me. I was driven to a gymnasium that had 500 bunks.

The gym was used to stage troops heading to Vietnam. I was admitted to what was referred to as the long term section of the gym. There were about a dozen of us off in the corner of this enormous building. I do not remember why all of the guys were there. Like me, there were several assigned to the hospital. They filled me in on the procedures. One guy was from Vietnam and was waiting for a flight back to the States. For the others I do not remember.

The deal was every morning I would take the shuttle over to the hospital and check in. Even though there was a mess hall nearby the gym, because I was assigned to the hospital, I had to eat all meals at the hospital. So, I was on the shuttle often. After I ate lunch at the hospital, I would take the shuttle to Clark's Enlisted Men's Club. I would hang around for several hours and drink a couple of San Miguel beers. I cannot believe I had very much money at that time, but then I think a beer cost something like 20 cents. Later I would head back to the gym.

After several days, I checked in and was told I had an appointment with a doctor. I sat around the hospital until my appointment. The doctor did his exam and announced that I did not have a hernia.

I said "What about sick call and the Field Hospital?"

He said "Look son, I am the guy who would do the cutting and I do not see a hernia". He then said that he felt I had a relaxed inguinal ring. He continued, "It will probably correct itself. If the lump pops out, push it back in".

The next morning I was next to the C-123 that was heading back to Vietnam to pick up the days wounded. I arrived back at the 3rd RRU and checked in with the Orderly Room. The clerk said so you finally got your hernia fixed. I explained the story to him. The First Sergeant was standing nearby. After hearing the story he said, "Fucking Air Force".

I left Vietnam in April of 1965 and the Army in 1967. For a period of time the lump would show up and I would push it back.

After the Army, I worked for the National Security Agency at Ft. Meade. In 1970 I was taking some vacation time and going home to Philadelphia for Thanksgiving. A few days before I headed to Philadelphia, my lump was back. When I got to Philly, I went to see the family doctor. He said "You have a hernia." He asked how long I had the lump. I told him the story and he just laughed. My father, a WW II vet, said I should go to the VA since it was military related. I told him I had health insurance. The next day I rode the #5 bus down Frankford Avenue to St. Mary's Hospital. I met with the surgeon and he diagnosed a hernia. I was admitted to the hospital. The next day the hernia was fixed.

Just as an aside, the evening of the surgery I went to use the bathroom. My penis and testicles were black. I pressed the call button and explained to a nurse. She said she would call a doctor and a while later, a woman doctor arrived. She looked and explained it was blood seepage. Then she said, "You are not going to lose your stuff; it will go away in a day or two". That it did. I have had no re-occurrences.

The Hermaphrodite
By Lee Bishop

"Hey, GI, you buy me Saigon Tea?" the bargirl asked me.

She said her name was Thuong. I was about to give her my usual, "take a hike" speech in Vietnamese, but stopped myself. Here I was, in downtown Pleiku, the only GI in the Snow Bar and a Vietnamese linguist at that. I figured I might as well play along and have some fun.

"Sure, I'll buy you a Saigon Tea."

"Oh, GI, you very nice, very handsome. Come! We sit in other room!"

Well, the other room was like a large living room might be back home in the States and, since there was no business to be transacted at the bar, all of the girls, about a dozen, joined us. Thuong and I sat down on a sofa with the other girls clustered behind her.

I had a cognac and coke while she quickly went through a number of Saigon Teas and kept complimenting me in English. But at the same time she kept up a string of comments in Vietnamese to her friends. *"Look at this stupid American. He spends his money on tea like it is the way to my heart. I can control him with just a look and a smile. He is so stupid! Of all the girls in this bar, I am the one he came to. You can all see why I am so popular."*

For my part, I kept a goofy smile on my face like I had no idea what she was talking about. I let the charade go on for about forty-five minutes as she reiterated her control over me and her superiority to the other girls. And the other girls were showing her deference, because she was obviously in charge.

Right after one particularly devastating shot at me that got lots of laughter from her friends, I turned to her and said in a loud voice so that everybody could hear, "Toi nghe noi la me cua may thich bu cat cua Viet Cong bi chet roi." ("I understand that your mother likes to suck off dead Viet Cong.")

This was followed by shocked silence from everyone. And then several things happened: About the worst thing one Vietnamese can say to another is to call them a monkey, and I had clearly made a monkey out of her. And the direct translation of American slang left her speechless. In a single sentence I had managed to offend her mother, herself, and the dead. Also she had invested a lot of time and effort into building a superior position above me and above her peers. And yet now she could not have lost face more quickly. The other bar girls responded to the situation by laughing at her, taking up my cause, and physically moving behind me on my side of the couch, patting me on the back, and giving me the appropriate level of adoration for a brilliant set up and sting.

In time, the bargirls of the Snow Bar and I became friends, and I would try to stop by and say hello whenever I could.

During one of these visits, I was sitting on a barstool at the end of the bar with the girls clustered around, when they told me that one of their co-workers was a hermaphrodite. I was sure that I misunderstood, but they kept repeating it, laughing all the while.

"Look! We show you!"

They called a young person who looked just like an ordinary girl into the room. They said her name was Mai and told her to flash her breasts… and very nice breasts they were. She kept rapidly lifting and dropping her blouse. But as she got worked up, the girls started laughing and pointing. I looked down and sure enough: there was a six-incher poking out very happily under the black silk of her trousers.

I was amazed and knew I had to share this one with my friends.

I got back to Engineers Hill in time to start my job as trick chief for the afternoon shift. There were nearly two dozen guys working as Vietnamese and French linguists plus our ditty boppers (Morse code intercept operators).

Once we got most of the day's traffic out of the way, I told my story. Of course, no one believed me so I offered to take them down and show them the next day. Everyone wanted to go.

So, early that afternoon, the Snow Bar had more daylight business at one time than they'd ever seen. We all ordered ba moui bas and went into the living room area with the girls. There was plenty of drinking and laughing and flirting when all of a sudden we heard, "Holy shit! Hoooly Shiiit!"

There was Ed Bluestein with Mai on his lap, and his hand wrapped around her major hard-on. Everyone just stared. They saw it was true after all and I was vindicated! Here was one person who was happy to be both a man and a woman!

We left the bar cheering and laughing and joking. Stuck in dangerous job a half a world away from everything we had grown up with, we shared a moment when we could forget all the mangle and madness of war and wonder at the strangeness of life itself.

A Little Rest & Recreation
Becomes a Lot of Escape & Evasion

By Lee Bishop

"Give me another one of those beers, Lee," Walt said

"Here ya go, Walt, but the party's almost over. There's less than a case left."

My friend grimaced and asked, "Wonder if this old gook knew he'd party more after he was buried than he ever did while he was alive?" We were sitting on the tomb of some unknown Vietnamese person.

Walt Smith was blonde, medium height, blue-eyed and heavily muscled. A real American Golden Boy. How a corporal in the 101st Airborne's elite Long Range Reconnaissance Patrol (LRRP) and a Vietnamese Linguist in America's elite Army Security Agency (ASA) became close is something even we hadn't figured out. We just enjoyed each other's company.

As usual, we were partying with a dozen other guys in the sand around North Air Field. This was a mile inland from Tuy Hoa and the coast of the South China Sea. Our favorite drinking spot was this solitary gravesite where we could get away from the craziness of the army and the madness of the war.

Vietnamese graves are interesting in that a low masonry wall surrounds the individual burial plot. We would sit on the wall, legs straight out into the sand, and trade stories, some from the war, but most from civilian life.

This grave was kind of a boundary at the foot of a sand hill. LRRP was at the top of the hill, and, since officially there were no Army Security Agency units in Vietnam, our "Radio Research Unit" sat at the bottom.

BANG!

More than one of us party people asked, "What the heck was that", or words to that effect, as we reached for our weapons.

"No problem," someone shouted, "Lt. Castleman just tripped over his own feet again. He was running with his .45 cocked because he heard us partying and thought Charlie had broken through the wire."

Walt said, "Hey man, let's go to my hooch. I've got almost half a bottle of vodka and some more beer up there."

And so we departed to start a night destined to live in ASA and LRRP infamy, or at least to stand out in my own memory.

We trudged on up the hill, entered Walt's hooch, and started on the vodka. That stuff must have been watered down because it disappeared pretty quickly. Then we started on the few beers he had.

Very carefully Walt placed two beers on the table.

"Thanks," I said.

"Don't mention it," Walt replied in a tone that sounded like he meant exactly that. Walt was very serious about his drinking. He flipped the chair around and sat down John Wayne style. A three-day patrol had left him with sunken hollows beneath his eyes and a patchwork of insect bites on his neck and face. It was sort-of an old man's face set on the compact and muscular body of a nineteen-year-old athlete's body.

Sweat rolled off his sun-reddened face as he threw his head back to drink. Most of the beer went pretty near his mouth. I laughed and he gave me an appraising look.

"So, all you do is sit in your hooch all day long and listen to your radio?"

I nodded. Walt laughed silently.

"Must be a real important part of the war effort."

"It is, Walt. I report directly to General Westmoreland. It's not my fault the fuckin' VC haven't learned to use radios yet. Anyhow, tell me about the 14 year-old you captured. You guys raid the Ho Chi Minh nursery or what?"

He ignored my question. And then, as if he'd suddenly discovered a great truth, Walt shouted, "This place really sucks!"

Of course he was right. North Field was a shit hole. The GP Medium I was living in was always hot, smelling of stale sweat. And the people I'd reported to were idiots. I'd run into some extraordinary officers in Vietnam, but the MI officers we reported to at this time were proof positive that "Military Intelligence" was an oxymoron.

What the hell was I doing here, anyway? After a year studying Vietnamese at Defense Language Institute, pretty much all I was picking up in my intercept work was static.

And then Walt had another idea.

"Let's go to Papa San's for some tiger piss," he said. It didn't take much urging on his part. Beer LaRue was the official French name. The bottle had a picture of a tiger on it, hence the moniker "tiger piss."

Now this new idea presented a logistical problem of considerable proportion; Papa San's was outside the wire on the west side of North Field. But Walt said he was 'pretty sure' he knew where the machine gun positions were, so we headed to the perimeter. I could just barely see him ahead of me, running easily in the dark, half couched with his arms at his sides.

Sonofabitch! The ground seemed to rise up and smack me in the face. No problem; I'd just tripped on something in the dark. Walt stopped.

"Nice going," he said sweetly.

"I don't do much of this shit when I'm sitting in my hooch," I said.

Walt laughed and helped me up. "You okay?"

"I'm pretty sure both my kneecaps are broken."

He pretended he was deeply concerned. "How's your dick?"

"Okay," I said.

"Good. Then you really have nothing to bitch about! Let's go."

We crawled into a drainage ditch and moved to within fifty meters of the first machine gun position. Walt said to wait, climbed out of the ditch, and moved to the position. The ditch started spinning, and I closed my eyes.

Walt came back and said, "I know where we can get through the wire."

"And we're not going to get shot, right?"

"Probably not," Walt said over his shoulder. I climbed out of the ditch and followed him. Looking back from now it seems crazy, but this was actually starting to feel like fun.

We crawled to three more foxholes to alert them that we were going through the fence to get a few brews.

Trip flares occasionally lighted up the sky, but that was typical so it was a pretty uneventful trip. We got a couple of what appeared to be quart bottles, found a comfortable place in the dunes, leaned back, and enjoyed the first cold beer we had consumed in nearly an hour. Unfortunately those were our last cold brews for a while because they were the last two that Papa San had.

Our important mission accomplished, we went back in the way we came out only to realize that our internal clocks were announcing that the party was just getting started. Walt asked; "Why don't we go for a little Rest and Recreation downtown, Lee?"

"Right, Walt. Where do we go for our evening passes? I'm sure they're going to let us bust curfew."

"No, man. We don't need any passes. We'll go out the north side of the perimeter the same way we went to Papa San's. Nobody's going to do anything. All we have to do is dodge the MPs."

"I don't know, Walt. People with security clearances aren't supposed to be as adventurous as you LRRPs. If we get caught, I'm going to be in deeper shit than you'll ever have to think about."

"To hell with that! Put on your party face, buddy, because we're going to get drunk and get happy all night long!"

Somewhere in that colloquy there must have been some magic words because I shook my head and said, "Let's do it to it, Walt!"

And we were off again.

Walt maneuvered us through the barbed wire and concertina as well as the machine gun positions so that we were able to exit the perimeter on the north side. Now we had to get across a blacktop road, through an area of tin hooches occupied by Vietnamese, and down a country lane about a mile to Tuy Hoa.

As we crossed the road we saw jeep headlights coming straight at us.

"MP's!" I yelled, and Walt and I ran into the hooch area, hoping to lose them. I got the bright idea of ducking into one of the hooches and was greeted by the timid stares of an entire Vietnamese family. Actually Walt and I were both fixed by those stares because he was right on my heels.

I quickly told the family that we were being chased by the military police and asked if they would help us. They got big smiles and told us to stay as long as we wanted…which wasn't very long because we were definitely wrapped up in the idea of more beer and meeting some ladies.

When the coast looked clear, we were off. The moon was bright and full so we could see pretty well as we walked down the dirt lane that led to Tuy Hoa and hopefully to the objects of our affections.

The lane into Tuy Hoa was dusty and rutted from the daily traffic of trucks and jeeps. On either side of the road the jungle edged in with tree branches bending far out over the side ditches filled with stubby cactus. In the daylight, from a distance, the jungle could be beautiful in endlessly intricate patterns of differing shades of green. Up close at night it was simply black.

Tuy Hoa was off limits at night so Walt and I pretty much had the road to ourselves. Still, we stayed close to the edge, remembering the sniper fire we'd experienced on other trips. There was a jungle trail that paralleled the road that was said to have considerable Viet Cong traffic.

I pointed that out to Walt.

"Every jungle trail in the whole damned country considerable gook traffic," he whispered back.

We came into town on the far west side. The lane we were on was bordered on the left by the backs of various shops and on the right by about a six-foot drop-off into what looked like sand and vegetation.

We heard a jeep coming up behind us.

"MP's!" Walt croaked in a whispered shout as he shoved me over the embankment and jumped after me.

"Oh, crap, man", I whispered loudly. "We're in a patch of cacti. This is killing me!" And then I started laughing.

"Be quiet, you dummy! We're going to get caught if you don't shut up. Don't move and don't say anything until the MP's are gone."

So we lay there, choking off our laughter, convulsing in silence, and wanting to scream, not breathing another word as the MPs' in their open jeep slowly drove by.

We struggled up the shifting sand of the embankment, wanting nothing more for the moment than to stop the pain. We pulled spines out of each other's backs and butts for several minutes, and then it was off to partake of the pleasures of the flesh.

Suddenly we didn't give a shit about the cacti, the snipers, or the MPs. We started laughing and talking out loud. This was our own private little battle, and no one else was invited.

"Except the whores," Walt solemnly reminded me. He was right. Whores were invited.

As we headed east down the road we fell in behind a Vietnamese girl. Walt whispered, "Boy, I'd like a little of that!"

As we got closer, it turned out to be a friend of mine named Huong. Now Huong was well known to a lot of the guys, but respected because she dated an ARVN named Beh who was assigned to work with us. Beh was a good guy, and he and the other ARVN support person, Vi, helped us through a lot of tight spots. However, while we were in the field at Phuc My, Beh told us that he was no longer dating Huong because she had been dating GIs and they tended to stretch out a girl's pussy.

The timing was bad because I had no more than said hello to the girl when Walt yelled; "MPs, run, man!" And we took off through the alleys. But they were really on us this time so we split up. I dodged into a couple of different stores with the same story I had used in Tin Town and got the same supportive reaction. After losing sight of the MP's, I circled back. I saw nothing of Walt, but Huong was still in the vicinity.

I told her what was going on so she took me to her grandparents' home. She said the MP's would be doing a house-to-house search for us. Her grandparents hid me under their bed until the search was over. I thanked everyone and meandered through town looking for Walt and downing a few Cognacs and Coca-Colas.

Now I couldn't find Walt, but I was feeling no pain. I was, however, still sober enough to know that I'd better get my tail back inside North Field before dawn, or I'd be living with some consequences that I did not want. Or the VC would nail me, and I wouldn't be living at all. So I started wandering back up the country lane toward Tin Town at a less than a steady pace.

Not far into my new quest, three schoolboys surrounded me and started yelling, "You teach me English! You teach me English!"

I said, "I can't boys. I've got to get back inside the compound, or I'm in big trouble."

They offered me a deal. "You come my house, teach English one hour, and we get you back inside. No problem."

At this point I was thinking, "Nothing from nothing leaves nothing, so what the hell."

"Okay, boys! I'm your man!" And off we went to their house.

After meeting mom, dad, an aunt, and grandpa and grandma, I sat down with a book the boys provided and gave them what I suspect was the worst English lesson of their lives.

But, true to their word, they escorted me back to the east side of North Field. By this point, it was a very dark and stormy night. The rains had started, and with the moon behind the low and heavy clouds and the big drops pelting down, it was impossibly dark.

I was facing an eight-foot high tornado fence, reinforced with a pyramid of concertina wire; big, round roles of razor wire set in a row three deep, topped by a row two deep, topped by a single row. I figured things weren't looking too good. At the same time, I couldn't see much more than ten yards in front of me and I knew the guards couldn't see any better. If I was going to get shot, there was a good chance it was coming here and now.

One of the boys whispered, "You come here, GI. Here is hole. You crawl through. Nobody see."

And, even in my inebriated stupor, I was thinking, "Jeez, I'm not even old enough to legally drink hard liquor yet, and here I'm probably going to die because of it!" But there were no viable alternatives and so in the black hole I went. It was a surprisingly easy crawl, and I was snug in my sleeping bag within ten minutes, never having received a single challenge from our guards.

With even young native kids knowing how to get into our supposedly secure position, I did have some questions about how protected we were. Of course, that was a question that I had to keep to myself, since I would have been forced to give the whole story and that would have gotten me court-marshaled.

Walt found me the next day and asked how I'd fared. I gave him a general run down and then asked, "Where did you disappear?"

"Oh, man, I thought I slipped them when I ducked up an alley. Except it dead-ended against a wall. The MP Jeep pulled up to block the only way out, and an MP captain got out with his .45 drawn and shouted; "Come out of there soldier! Right now!"

"I figured he knew what he was doing, so I walked out, cold-cocked the SOB and took off running like the devil. I found an all-night pleasure house and left part of my brains there on the sheets. Man, you should have stayed with me. I had a hell of a good time, Lee!"

So now you see how a little Rest and Recreation (R&R) became a lot of Escape and Evasion (E&E).

Several months later, after some training up in Phu Bai near the DMZ I heard that Walt bought it in a firefight. Losing friends was always difficult for me, but this one really hurt. Some of the guys made it through and some did not, it was all the luck of the draw. But thinking back, I'll never forget my good friend Walt, and I'm glad we had our little adventure together.

Chapter 3
Special Events

Orange is Not a Primary Color
By Robert Flanagan

One sunny afternoon in the late summer of 1964, shortly after I'd been promoted to SSG and become NCOIC of the Air Section, 3rd RRU, I returned from an ARDF mission in the Tay Ninh area. My pilot, WO1 Ray McNew, taxied us near the end of our designated Army ramp area as our bird was scheduled for maintenance. We grabbed flight bag, mission bag, helmets and weapons and ambled down the ramp toward 'The Hangar' for debrief. Most of the 68th Armed Helicopter Company's Hueys were afield and the ramp was relatively untenanted ... with one surprising exception: a battered, down-at-the-heels, Air Force C-123 appeared to have collapsed in disgust on the verge of the taxiway-ramp. It sat, soaking up late sun and I saw—or maybe just imagined—a cloud of flies hovering over the dejected hulk.

An Air Force Tech Sergeant wearing a beanie sat on a 55-gallon drum adjacent to the lowered rear ramp of the bird. The drum, mustard in color, had three pink bands circumferentially.

I said to McNew, "Sir, I'll catch up with you. I want to take a second." He acknowledged my plea by ignoring me and kept moving. I walked up to the Tech Sergeant, an Air Commando, and saw prominently displayed on the chest of his baggy gray flight suit a patch that read Ranch Hand.

The Sergeant watched me with a baleful glare as I approached. "Hi," I said, and being he was a Zoomie, tweaked. "Where do you drop your bedroll?"

"Huh?"

Nodding at his patch, I asked, "What spread you riding for, cowpoke?" I think I grinned to take off the edge.

As if not noticing the patch before, he glanced down and said, "That's our operation name."

"Your bird got maintenance problems?" I asked, looking down on a pool of viscous green-and-orange liquid muck that did not seem to mix, but showed overtones of both colors; a strong chemical smell, astringent and slightly sweet, overhung the derelict.

"Just some down-time." He seemed reluctant to talk and looked away from me, across the field where VNAF A-1Es were making touch-and-go. Looking back along the length of the airframe, I counted nine bullet holes. The paint was flaked away about the perforations, the aluminum skin showing through, emphasizing their distinctive wrongness in the scheme of good airplane protocol. I thought about the long line at the sheet-metal airframe shop.

I recalled the cells of 123s we'd occasionally encounter while on mission. We flew low to get good DF lines with minimal flight disruption, but the cargo craft flew LOW! Looking down from our meager altitude of 2,500 to 3,000 feet, I estimated them to be doing their thing at 500-900 ft. They reminded me of times as a child in the Mississippi Delta cotton country, when I was privileged to watch a daredevil roaring down the length of long cotton rows, to rise precipitately at the end, roll and twist his small craft with minimal waste of airspeed, to return on the opposite tack some rows over, all the while casting a white cloud over the budding cotton shoots.

I looked into the well of the cargo hold; there were no bucket seats, as with 123s used for transport. More barrels, bearing a rainbow of

stripes—green, pink, white, orange, blue, purple—
were lashed to the walls of the craft. "What's your
mission?" I asked the circumstances of multi-
colored disarray strange to me.

"Classified. Classified mission." He didn't
look at me.

"Classified?" I said. "Hell, Sarge, see that
little tinker toy over there I just got out of, that
Beaver with all the antennas on it? Now that's
classified. Seriously classified. But you're a GI, like
me; we could exchange notes."

His eyes came around slowly as a turtle's
extended neck-head movement. He didn't reply.

"What's that orange gunk pooling at the
ramp?" I asked.

"Kool-Aid. Orange Kool-Aid. We took a
round in the galley." He said it with no sense of
humor; someone might have believed him. "So,
fella, you can tell me about that Beaver's mission,
huh?" he questioned.

"Sorry, classified," I said curtly.

"Shove off, Sarge. You're on restricted turf."

I rambled off after Mr. McNew, thinking, Kool-
Aid, my ass!

Army 1-5-1, MAYDAY!
By Robert Flanagan

Dan Bonfield and another errant 058 of the Air Section, 3rd RRU, found drinking *Ba Moue Ba* to be a pleasing enterprise after tense hours aloft, thus they did quite a lot of it. In early January, 1965, the pair returned late at night to Tan Son Nhut from downtown Saigon, exited their taxi at 100-P alley, and made it past the newbie Air Policeman at Gate 2 without incident. Hiring a dong cart, they were transported to the far back corner of TSN with unremarkable speed, the steel-banded legs of the cart man pistoning tiredly. About 100 yards from the gate into Davis Station, they called the driver quietly to a halt. They had a plan!

The two happy inebriants thought it would be all shits-and-giggles to attack the station and scare the guard. Not altogether prohibitive as a wild-and-crazy thing to do on a wet night ... but they hadn't a clue as to who the guard on the gate was. When they piled off the cart, stumbling and giggling, more sober minds might have predicted the guard must have heard their antics: it was a still night, no aircraft traffic. An occasional dog barked from the ARVN family Quonset hut quarters, but the guard, leaning on the wall in the shadow of the Post Office, was a mystery. Still, the opportunity was there and the two were up for it.

They crept forward, using the shallow ditch that lined the asphalt road to reduce their profile. At something less than 30 yards, using some jointly agreed signal, both leapt from the ditch and raced down the center of the road screaming, "Amelican, you die. Western Imperialist dogs, Ho Chi Minh send gleetings," and other retro distorted dialogues from bad WWII movies. It is unlikely they really imagined they would fool anyone.

Where the road they used crossed the road which ran directly into the gate of Davis Station, and opposite the P.O. guard post, an ARVN Ranger Training unit occupied the ground. Rangers, whatever their nationalities, are known as preemptive soldiers. Preparing for the 'What ifs?' the rangers had constructed a sandbagged machine gun emplacement on the corner of their turf on street's edge. It was merely a corral of sandbags about three feet high with no overhead protection; in the center of the emplacement they had sunk into the ground a four-feet-long, three-inch-diameter steel pipe with a slightly flanged top end. This precaution was so that, when under attack, rangers might race to the spot with a .30 cal. Browning and slam it down on the top of the pipe, using it as they would a tripod mount, but already in place and very stable.

As the two besotted ops screamed down the road, their minds must have, in the exertion, cleared enough to allow them to hear the loud, ominous clatter as the gate guard cranked a round into his M-14 chamber. An instant aversion as the guard stepped forward, menacingly, into the street and the two made for the ranger gun pit. The other Op— I've forgotten who he was—sailed majestically over the top of the sandbags and knocked himself unconscious against the back wall. Dan tried with all his might, but in his half-gainer-like stretch for safety, he found he was unable to bend the steel pole with his right knee, though he gave it a monumental try. The pole remained impervious; Bonfield screamed in pain, enough so that after some satisfactorily complacent non-intervention, the gate guard, laughingly—invoking a small pity— called the CQ and asked him to send for the meat wagon.

Dan was in the Cholon Navy Hospital for a few days, then on light duty on crutches for a couple of weeks. His knee, initially the size of a watermelon, and colored much the same, had subsided gradually, and was almost down to human proportions when, on January 18th, he awoke, his leg still sore, but with a startling realization, overriding his discomfort. He dressed, skipped chow, and hobbled down to the motor pool where he inveigled a buddy to give him a ride to The Hangar where the ARDF Section NCOIC hung his hat.

Bonfield was still on the cane when he approached my desk where I was struggling with mysterious alchemy to make too few planes and too few ops stretch fit too many missions.   "Sarge," he said, "you gotta help me."

"Whacha need, Dan. I'm busy." Always something.

"Sarge, I gotta get some flights in. Soon. I don't have any days this month to cover my Hostile Fire Pay creds. This leg's kept me outta the bird." He had the look of a chastened Chesapeake Bay retriever. "Can you put me on a flight?"

At the time, the Air Section was one of a very few assignments in the RRU which, because of their regular exposure to enemy fire during low-level DF work, qualified the pilots and ops who flew missions to draw Hostile Fire Pay, a $65-a-month stipend that bought a lot of beer. And once accustomed to the extra pay, it was grievously hard to give it up.

But the reg governing award of this extra pay demanded exposure to enemy fire on six or more days in each calendar month, ergo six mission flight days; flying a mission was presumptive exposure to fire. Dan, courtesy of his *Ba Muoi Ba* catastrophe, didn't have any days.

I looked down, noted the cane, the still-swollen fatigue pants about the knee. "You're still on light duty; you know I can't do that." I shook him off and returned to the unequal math.

"C'mon, Sarge. I don't need this cane. Look—" he tossed it aside and tried to cover the grimace when that leg came down solidly on the floor. "If I can get off the light duty list, can you give me a flight? Right away?"

"Big if, Dan. And since we're short of everything but targets, and I'm already tightly scheduled, I'd have to make a shift. Guys already got their heads around the roster."

From the corner of my eye, I saw him slump dejectedly. I hesitated ... Dan was a good Op, and I was up to my gaiters in 'gators, so I made one of life's supremely silly decisions. "Oh-seven-thirty, tomorrow. The nineteenth. Take my early flight." I rationalized my action on the basis of a need to catch up on onerous paperwork and I already had my six days in. "But, you gotta be able to walk, Dan. You gotta get off the Doc's light duty status and walk here from the hooch—no bumming a ride," as I looked at the disinterested driver waiting for him "... and no cane. No walker, crutches, gurney. None of that shit. Walk here, get your gear, walk to the flight line, and mount the bird yourself. The pilot is, let me see, Mr. McNew. Any last-minute brief info at oh-seven hundred."

"Thanks, Sarge. And don't worry. I'm ready to fly. Besides, we've never lost a bird."

"You'd better hope," I said with a gimpy eye. "That leg's not good yet, and if you went down you'd be a goner. I'm hanging my ass out on this one, Dan. Don't let me down."

By the time I reached The Hangar the following morning, McNew/Bonfield's flight had launched. I never thought about it; I had three other flights out in various regions, as well as the VNAF ARDF bird which our shop also tasked, but the VNAF flight might be anywhere from Ca Mau

Peninsula to Frankfurt. Mission tasking was hit-and-miss with the Vietnamese DF-ers.

Along about 1000 hours, SSG John Moore, T/A NCOIC, leaned around his door and called, "Flanagan ... ringy-dingy."

I picked up the phone without anticipation. "Air, Sergeant Flanagan ..."

"Sarge, this is Lieutenant Lee, S.A.R."

"Yes? S.A.R., sir?" I'd had no occasion to know the term.

"Search and Rescue. You guys have an Army bird, tail number suffix one-five-one, operating up around Black Virgin Mountain?"

"Uh, Lieutenant. What's the purpose of your question? Our flights are classified; I can't discuss their whereabouts."

"Well, just thought somebody might be interested. We just got a MAYDAY from a bird somewhere in that region; he squawked as 'Army one-five-one,' and declared a full emergency. Other chatter picked up after the initial call indicated a loss of power and that the pilot was going to attempt a dead-stick on an unregulated field."

I couldn't reply.

"Know anything about that. Where he might be? If he's yours."

"Uhh, yes, lieutenant, we have a Beaver, a RU-6, with that tail number trailer. He might be in that region; we don't log flight plans primarily because we don't know where our missions might lead us."

"Well, his apparently led him into trouble. If you can pin down a possible location of this so-called 'field' he might try to land on, call me back. Tiger seven-seven-zero. It may help S.A.R. locate them."

I hung up, thinking furiously: I knew at least three fields in that area, sod landing strips on various French-owned rubber tree plantations, used by the owners or overseers to get back and forth to Saigon and elsewhere without having to run VC roadblocks. I'm dialing while thinking.

When the clerk in the Air Section shop had passed the phone to Captain Carlisle, I said, "Sir, Sergeant Flanagan. One-five-one's down in the Michelin."

There was no dead time. Carlisle, an older Captain, infantry Platoon Leader in Korea, could read a scenario in a heartbeat. "Who's the pilot?"

"McNew. Bonfield's his Op." McNew was a relatively new Warrant Officer, one just months out of flight school. Ostensibly, perhaps, not the pilot you'd choose to handle an emergency. But, who's to choose?

"Bonfield's back on status?" There was now an edge to his voice.

"Uh, yes sir. We can talk about that." Oh, Jesus, I thought; would we ever talk about it.

"Any pilots about the shop over there?" he asked.

"Uhh, I saw Mr. Martens up in S-1 a few minutes ago."

"Get him; I'll alert the line to get a bird ready. Once you're in the air, redirect any other 3RRU aircraft you can reach to join on you. Do you know where they might be?" He knew that, by tasking, I would know where they were supposed to be. Maybe. If the wind hadn't shifted. If a new, aberrant VC target signal hadn't suddenly arisen into the ether to draw Bonfield's attention elsewhere. If McNew hadn't decided to fly to Vientiane.

"I know three possible sites. We'll check all three. Out." I dropped the phone into its greasy case, folded the onionskin copy of Dan's mission sked into my pocket, grabbed my flight helmet off my personal nail in the wall, took down an M-14 from the common use rack and headed up the hall.

Stuck my head in S-1 and interrupted one of Mr.
Welch's interminable drunk jokes, motioned CW3
Martens into the hallway, and explained the
situation. An old hand at crises, Martens grabbed
what he needed—pistol belt, helmet, flight bag—
and we sprinted for the door.

It was but a few hundred feet out the door,
around the edge of a maintenance hangar, to the
Army flight line where we spotted an SP-5 waving
us toward an RU-8D, twin engine DF bird. I raised
the right hatch, clambered in while the chief did a
runaround preflight. We got emergency clearance,
jumped the takeoff line, cutting into the active '2-5',
the main airstrip at a bearing of 250 degrees, under
prevailing winds; when utilized from the opposite
direction, the same strip of concrete was nominated,
'0-7,' or a bearing of the reciprocal 70 degrees. We
were airborne inside five minutes.

We were still climbing when we heard
another of 3RRU's ARDF flights calling for landing
instructions. I looked at the onionskin: it was
Lieutenant Hall, inbound from a mission toward
Vung Tau. I looked at Martens with the question; he
nodded. I keyed the mike and when I'd made
contact, relayed the Captain's instructions that if he
had sufficient fuel, Hall was to join us for an
emergency search. He conceded he had some fuel
to burn and asked for our location. Mr. Martens
chimed in with that, and it was only a few minutes
until I spotted the other U-8 plowing toward us at
max thrust. Hall was always up for something new.

The two planes in tandem strained on toward
the prominent landmark ahead, the 3,268 feet high,
pimple-shaped extinct volcano Nui Ba Dinh, 'Black
Virgin Mountain,' near Tay Ninh in northwest III
Corps. I'd reviewed the three possible sod strips I
thought McNew might have targeted and decided
we might as well go for the most accessible; it was

a dartboard exercise at best, but if McNew'd had a choice, it's the one he would have chosen.

The smudge that was Tay Ninh was growing distinct in the windscreen when there was a sudden loud crackling of radio transmission: "Break! Break! All aircraft in the vicinity of Tay Ninh, Dau Tieng, Tri Tam and Nui Ba Dinh, vacate the area for emergency recovery operations. All aircraft, vicinity Tay Ninh, vacate the region. S.A.R. operations commencing."

"That's our invitation to go home," Chief Martens said, keying the intercom. "Army two-seven-six, this is Brown Shoes eight-oh-eight; you copy? Over." he broadcast to Lieutenant Hall.

"Two seven six, copy. Out."

"Eight-oh-eight. Out." As we banked into a heading to take us back to Tan Son Nhut, Mr. Martens broke into an old Southern gospel hymn I recalled from childhood. "Sometimes I feel like a motherless child. Sometimes I..." The song had never made sense to me before.

We were met inside The Hangar by Captain Carlisle and the S-2, a young 2LT, along with a host of curious hangers-on, everyone shouting questions. We had no answers beyond the S.A.R. broadcast. A clerk called out from the S-1 shop, saying S.A.R. was on the line. Carlisle took it. He came back in, nodded at me and said, "Good guess, Sarge. They went in on the strip you picked on the edge of the Michelin. They've got both men out, but the bird's still there. Badly broken. What about the radios?" he looked at me.

"Standard ops for landings and all emergencies: zero out the dials, shut down power; eat your rice paper sked sheet." That was it. We'd no precedent. McNew and Bonfield were doing an O.J.T.

Carlisle motioned me and the S-2 to accompany him to a quiet corner. "Flanagan, you

99

know everything the Op's supposed to be carrying. Barry," he said to the wide-eyed S-2, "get a camera. I'll get a senior mechanic and get us a ride. We have to go out to the bird. Check it out. They might or might not be able to recover it."

I knew the camera would serve three purposes: to portray conditions vis-a-vis security, pictures for an aviation accident board, and bragging rights in the bar. The S-2 looked as if the notion of flying toward a trouble spot was alien to him. But by the time we rounded the hangar to the ramp, a UH-1 was settling onto the apron, a slick with only one door gunner. The four of us clambered aboard and found straps to hang onto; there were no side canvas seats; this bird didn't do limousine service often.

At the end of the sod airstrip, as we banked and made an approach circuit, I looked down and saw a suspiciously dismembered Beaver scattered through the rubber trees at the end of the strip. We slipped onto the ground between two circling Huey gunships that were orbiting the crash site. We un-assed the Huey and began our tourist act.

The fuselage of 151 lay on its port side, the port landing strut and wheel torn away. The wings, ripped from the fuselage, both lay some distance away, one on either side, within adjacent rows of rubber trees. The tail section looked relatively free from grief, but in the circumstances, I thought it mattered little. Initially, none of us seemed to notice, but then the door gunner from our Huey — who had dismounted his M-60 from the bungee cord and brought it with him to accompany us—shouted, pointing up the same row of trees in which the fuselage lay, and there it was, some 100 feet away: the engine. Ripped free from the firewall-cowling, it had made conspicuous tracks where it tumbled forward of the down site.

We heard a few rounds of AK fire whistle through the rubber trees, high above us, and an attack lacking seriousness, especially when one of the circling Huey's made a firing pass along the edge of the strip. The S-2 took pictures at the Captain's direction, the aircraft mechanic pored over the wreckage, and we were visited by another Huey-load of straphangers, including the 3RRU Commander, LTC Thomas Owen. But the critical thing is that we found the rack of two R-390 HF radio receivers and a generator, thrown clear of the aircraft, and the frequency dials were all set on zero. Bonfield had done his job! Neither the radios nor the bird itself would ever function again as designed, but we had our men back and we'd not suffered a security compromise. After maybe 45 minutes, we boarded our taxi and flew home. So to speak.

That was the first aircraft loss the 3rd RRU suffered. Later, with the 3rd RRU Air Section morphed into the 224th Aviation Battalion, with four ARDF companies and the CRAZY CAT RP-2V collection platforms, we would lose others. Some would bear a heavy cost, as in the loss of all five men from the 138th Aviation Company in March, 1971. But that was all in the murky future when 151 went down in the Michelin plantation.

Bonfield had a broken collar bone and possibly some broken ribs. He was thoroughly black and blue all over from the force of the equipment rack falling on him as well as the shock of the crash. It was only hours after his initial examination in the hospital that they discovered he also had a bullet lodged up under a shoulder blade. Mister McNew was lacerated and severely bruised over his entire chest from the crash into the yoke upon impact. Both lived, were awarded Purple Hearts as a result of wounds sustained in enemy action, occasioned by the ground fire directed at the 'glider' as it was making a dead-stick approach to the sod field, and in taking evasive action to avoid the fire, overshot the strip and crashed into the rubber trees.

It was unsettling for me to think, later, that it was really my scheduled flight which went down. It was only Bonfield's financial penury that put him in the back seat in my stead. I never saw him to talk to after he was med-evaced, never saw him again, but wondered if, in some anguished late night reminiscences, he had blamed me for his fate. Ray McNew and I talked with wonder and humor, as lately as a couple of weeks before his death in 2013, about that weird and unlikely sequence of events. The relief of survival puts a soft, fuzzy patina over the most outrageous goings-on.

Warp Zone
By Robert Flanagan

I recall it was late in the day, though plenty of sunlight yet to go. As an A-Slash at White Birch manual Morse operations, a Spec-5 with aspirations, I'd ridden the duty truck back to Davis Station after a day shift and in the heat and humidity thought a shower might be a good use of my time. But as I didn't overwhelmingly reek yet, and I knew the White Elephant— our dismissive and derogatory reference to the CO's pet project, a large compendium of white tanks and gauges and switches and pipes which were, in aggregate, designated a hot water heater—did not bear up to its billing. There was no hot water. There was only tepid water, and though slightly better than the alternative, I couldn't help remembering taking salt-water showers in my days aboard the USS Antietam (CVS-36) during Marine service six years before: like tepid, a salt-water shower left you feeling cleaner ... but unclean. But this was, after all, a war zone. Warp Zone, John Moore called it, parodying some BBC utterance.

Off the truck, I stopped outside the mess hall—too early yet for chow—and was talking with someone about something: all very nebulous, when, almost involuntarily I began sniffing the air. Something odd. Something unpleasant. Something ... threatening wafted its way across the Davis Station spaces, borne on the stagnant, waste-redolent air of summer Viet Nam.

Gas! Gas attack! Someone down the company street decided for me, screaming in a fear-filled voice. I quickly realized he was right. The door of the Orderly Room smacked open and the CQ, just coming on duty for the evening, emerged and grabbed the designated steel bar and began clanging on a 155-mm spent artillery shell suspended there for use as an alarm for gas or air attack. He screamed his choice of Gas! Whanging away at the shell.

As in a tortoise colony, heads began to pop out of hooches; guys strained their necks around doorframes and out of latrine stalls to locate the source of bother. In a constant and increasingly pungent onset of acidity and a stinging sensation, the gas soon flooded all spaces in the compound. More screen doors crashed back as those assailed fled outside seeking respite, only to find there the source of their fear. Some, reacting to belated recognition, ran back inside seeking gas masks, likely issued upon arrival, stored in a foot locker, and never tried on, fitted, or checked for efficiency. That surmise proved itself later as troops with Martian faces choked and vomited and strangled inside the questionable safety of the devices.

The First Sergeant emerged from the Orderly Room, effecting calm and control until he began choking and, through tear-flooded eyes, sought his own mask—which, it turned out, he'd allowed the house girl to put away and he hadn't a clue where. Another Senior NCO, the Master Sergeant Maintenance NCO, an old soldier who'd seen combat in North Africa, Italy, France and Austria, with an encore in Korea, was struggling to get his own mask in place. In his late 40s, overweight, too many Happy Hours in the NCO Club behind him, still he seemed more concerned to spread the alarm that others might not suffer. "Get outside! Gas attack! Grab your ... masks. Get ... them on!"

He faded as his aged lungs, up to the challenge of the millions of Camels he'd smoked, were not up to the debilitating gas. He began to stumble, his warnings winding down in synch with his mechanical-toy movement. He collapsed on the cinder track about the hooches. Panic-stricken troops ran around him.

The CQ, who had no mask, maintained his role of professional alarmist for as long as he physically could, holding a hand tightly over his nose and mouth, alternately sneezing and vomiting to the side, and continuing the din on the shell.

The XO ran out the rear of the orderly shack and disappeared toward the motor pool. The CO was off somewhere, unavailable to share our games. My own mask didn't fit properly, and I felt the burning in my eyes and throat that, I hoped, signified mere tear gas, but also sensed the queasiness that argued for a vomiting compound. The fear was that these common harassing gas symptoms might be masking far deadlier components. It was an indoctrinated concern.

A general recognition of the threat drove men everywhere to seek, first, their weapon or a weapon, and second, shelter from attack. Early in the war, we had ridiculously insubstantial bunkers; hardly anything to warrant the term. Along the back of hooches backing onto the fence separating 3RRU from the flight line, a long ditch zigzagged in a faux promise of safety. A double row of sad, often flat, expired sandbags surmounted the rim of the ditch; it didn't even qualify for the nominative of trench. And after its initial construction—the result of an overly prescient imagination on the part of a previous First Sergeant—it was largely ignored, the edges crumbling and eroding, often containing the detritus of crowded living. The stench from this cartoon Maginot Line indicated it was closer to the hooches than was the latrine. It did not inspire a sense of security.

No one could locate the armorer; normally no one would care, but he had the keys to the arms room. Another set was, in theory, accessible by the CQ, but this safety factor was considered such a risky proposition that the nervous Captain CO, who feared enlisted men, kept the second set of keys locked in the company safe. And as both CO and XO were out of the picture, no one had the combination. Sibel, I think it was, ran to the motor pool for a tool of some kind to pry open the steel door of the armory. Personal weapons, mostly Swedish-K's and .357 magnum revolvers, a few scattered PPK, Luger and other favorites, appeared in the hands of far too many troops. This sudden flood of weapons in the hands of nervous troops, I judged, was a greater threat than the ambiguous gas.

Panic was the order of the day. There was no leadership beyond a couple of sergeants who tried to find meaning in falling back on common discipline; but discipline was hard to instill while threatened with disaster.

In one of those curious hiccups of calm that can occur in the midst of the most enervating chaos, suddenly, in an instant, no one was screaming or yelling; the alarm shell got a break; silence was an empty promise. In the sudden void could be heard the tantalizing sound of laughter. Not giggles and chuckles, but belly-broaching laughter. It floated across the compound of stricken soldiers, seemingly originating beyond the corrugated metal fence - on the airfield side of the divider. Over in the area of the 68th Armed Helicopter Company's chopper ramp.

Everything stopped. The sound of merriment was too incongruous for this scenario; it upset the balance of fear and anger and drove a peak of curiosity. A Warrant Officer, one of our 3RRU airplane drivers, moved toward the fence with his face in a scarf, crossing the small wooden span over the drainage ditch. He jerked open the door in the tall fence and disappeared.

He was back in a few minutes, his face bright red. I was still scrambling to adjust my own mask, and thought he'd gotten a particularly heavy dose of the gas. But it wasn't that; Mr. Welch was pissed! He stopped in the company street and informed any and all near enough to hear that the 68th had been loading crates of CS gas into a Huey and, during some playful grab ass, had dropped a case of the irritant, rupturing two canisters. CS is an insidious devil's brew of tear gas and a vomiting agent, used for crowd control and as a disarming tool; it is not, on its own, deadly. We lived, then in 1964, in the time of the coups, one Vietnamese general after another seeking to elevate himself above his fellow officers and politicians to rule the kingdom; and the followers of these erstwhile 'Kings-of-the-Hill' often gathered in threatening crowds in downtown Saigon, sometimes about the US Embassy, depending upon how they viewed America's relations to their aspiring hero. Huey-delivered CS was a handy deterrent. We'd known of its use, but had no experience with it.

Outrage was the order of the day, as the word spread and troops emerged from the most unlikely bolt holes. Personal weapons vanished. The panic index reflected a serious decline within short minutes. But outrage remained for the arrogance of the chopper troops who'd been so careless, and then when complicit in such an accident, didn't bother to spread an alarm. Several of the door gunners and crew chiefs who went to their tasks daily, walking from and to their hooches down the road, transiting 3RRU turf to pass through the fence to their aircraft, were barred from what had become a staple of their lives: a post-mission brew in the EM club on Davis Station. That'll make 'em think twice next time, seemed to be the notion.

The old Master Sergeant went away; he didn't die of the gas, but was so disabled by the

incident he was medically retired. The CO
mandated changes in arms room etiquette; we
presumed some ordnance-handling criteria were
examined and modified beyond the fence. But life
slipped back into the same wretched pattern as
previously. And summer was just beginning.

Attacks, Shots & Casualties
By Don Collins

We had a helicopter pad across the road from us, containing all of the Brigade's gunships. The other side of the pad was the perimeter, and a large minefield. On the other side of the valley was a South Vietnamese fort, and the town of Bien Hoa. At one end of the valley was the airbase. It was felt that the Viet Cong would not be dumb enough to enter a valley that had military installations on three sides, and the gunships were considered secure.

One night they came through the wire, and blew up the gunships with claymores. As soon as the shooting started, the XO fixed his bayonet, and went charging out the door, bayoneting the company clerk, who was running in the door. Before it was over, shrapnel would wound three other sergeants. In addition, a truck coming back from town, loaded with 173rd soldiers on pass, was hit with a claymore, killing four, and wounding thirteen.

Once the shooting died down I had a chance to look around, and noticed that Van Walter, the crypto repairman, was wearing white pants. Thinking that a little odd, I looked again, and realized that he was wearing his helmet, flak jacket, ammo belt, weapon, and absolutely nothing else. It turned out that he had been taking a shower when the attack started. At least he had his priorities in order.

I don't remember how many gunships had been on the pad, but only two survived the attack, and one of those was heavily damaged. The gunships were never parked there again.

On another night, we were sitting in the club watching TV, and the whole room just lit up like brilliant sunlight. As we turned around, the sound wave of a huge explosion hit us. The VC had blown up the Long Binh ammo dump. There was a column of fire about 1,000 feet in the air. We were about five miles away as the crow flies, and could read a small print paperback book by the light of the fire.

Over at the 303rd Battalion things weren't going well. They were near the ammo dump, and the explosion blew down two walls of the communications center, sweeping every piece of paper out into a field. When things settled down they had everyone in the headquarters unit out in the field with flashlights, trying to find all the classified documents that were blowing in the wind.

The Long Binh ammo dump was blown up two more times while I was in Bien Hoa, but the first time was the most spectacular.

After about ten months, in early February, 1967, I was transferred to the 409th Radio Research Detachment, supporting the 11th Armored Cavalry Regiment, and based at Long Giao, about 20 miles east of Xuan Loc. They were rather unique for an ASA unit, because they operated out of modified Armored Cavalry Assault Vehicles (ACAVs). The tracks had quite a bit of firepower, and they operated pretty boldly for an ASA unit.

One day, one of the ACAVs was operating out of a Special Forces camp when one of the Special Forces patrols came under attack near the camp. The 409th guys saddled-up, went to their aid, and severely pounded the small Viet Cong unit, allowing the Special Forces patrol to casually stroll home. This is the only time I have ever heard of an ASA unit attacking something.

At one point they adopted a German Shepherd. It was an American K9 that had been captured by the VC. The 409th found it during an operation, and kept it. The VC must have really mistreated it, because it hated VC with a burning passion. You couldn't ask for a better watch dog, if there was a VC within 50 yards the dog would go nuts. Eventually some brass saw the dog, and reclaimed it for the MP's.

While I was there battalion Headquarters decided that we weren't having enough Article 15's and ordered the CO to have more. The CO was kind of PO'd about it because the reason we weren't having Article 15's was that no one was screwing-up. He finally made up an alphabetical roster of everyone SP-4 and below. If there weren't any Article 15's in a given month, the CO picked the next name on the list, and gave him an Article 15 for whatever he could think up.

It so happened that the first one was a SP-4 whose major crime was that he had a name beginning with A He broke his thumb cocking a .50 cal machine gun during an ambush, and got an Article 15 for Destruction of Government Property, namely, his thumb.

Seeking Cinderella
By Robert Flanagan

Vietnam: June, 1969. Benford was strange. I don't mean his mammary fixation; nothing strange about that. But I worked with him on the ground, flew with him on occasion: he was strange with a capital W.

His diversity likely had its genesis in nurturing by a family of devout Southern Batiste extremists who, after the shock of Benford's unheralded birth, relocated in the early 'fifties from the Alabama highlands to L.A. Adopting a reactionary Beat life force in the transition, adhering religiously to drug devotions, they sought the Haight-Ashbury motherland for three years before learning they were four hundred miles too far south, in the wrong city. Directionally challenged, incensed at the lack of civility among their co-devotionalists who had left them in ignorance, they became overnight converts to Hollywood Buddhism, rebelled at its banality, tried a fling at Santería, and ultimately became Republicans.

But what happened after the mortar attack, though also strange, was somehow . . . beautiful; a love story of sorts.

The day before the attack that night, I'd been on the manifest to fly, but when I showed up at briefing there was a note on the bulletin board: "WO1 Brenner. Stand down. Wallings will take Controller today, 6/22. See Major Fitz at 1300. CW2 Winter." A matter of little concern, it eventuated, but curious at the time; one was almost never scrubbed from a flight that late. But Winter was running on fumes then. He was just back from Hawaii where he'd pulled R&R with Nickie, and there were lots of signs that it was not a watershed of connubial bliss. I don't remember what happened in the meeting with Fitz, but I was in my bunk later when I felt the incoming. Must have been about 2130.

Mortars. Heavy mortars, probably 82mm, impacting across the strip in the Air Force compound. Lousy gunners: they obviously were targeting aircraft, but were off by a good three hundred meters, the rounds impacting in the Air Force compound. After some time, when the incoming had ceased and the All Clear sounded, after the goddamned Koreans had opened fire with the 155s, whoever was available from CRAZY CAT commandeered vehicles and headed across the strip to give a hand in the Air Force area. Might as well; counter-fire would go on the rest of the night. Thanks to our allies, the gunners of the White Horse Division, even one round of incoming meant the end of sleep as we knew it.

Benford jumped in the Jeep with us. First thing I ever saw him step up for.

The in-country Flight Ops building was nothing but smoking splinters. There had been an Airman clerk on duty, but no sign of him; only some bloody rags. The latrine next door was gone too, leaving behind only the stench. One round had impacted on the Air Force nurses' quarters, across the tarmac. Air Police all over the place, a number of Air Force and Army officers, and for some reason assorted Navy EM. Probably from the Sea Bees billeted next to us. The standard Mongolian Maypole Dance/Cluster Fuck!

The Jeep I was in with Benford wound up at the nurses' hooch. Just the one round had found a target here, and though many of the girls were off duty and might have been here and conceivably hit, most of them were at the O-Club at a wing-ding for a visiting strap-hanger. Only one nurse was slightly wounded. The medics were working on her, giving more hands on care than appeared necessary.

There wasn't even much structural damage to her hooch, but furnishings and personal gear were blown everywhere. I was reconnoitering a dark corner and beneath residue when I heard Benford exclaim: "Whoa!"

We all gathered on him, expecting more casualties. There was no one. Benford, resplendent in fuchsia-colored swim trunks, a Hawaiian shirt, and Ho Chi Minh sandals, was holding before him, in awe, a woman's brassiere. A large, woman's bra. A large woman's bra. Or, as Benford read it, a woman's *large* bra. He was mesmerized. Turning it over in his hand carefully, as if gently manipulating the bounty for which it was designed, he hesitated, then read out in hushed tones: "Forty-four dee." There was stunned silence.

Then a reverent, hushed murmur from 360 degrees: "Jeez-us! Forty-four DELTAs."

It was only a moment before critical intelligence was sought.

"Whose?" Tantos, despite his vow of warp zone abstinence to his Chiricahua Apache wife, was not unaffected to the point of not asking; but Benford led the pack. And at that moment was initiated an inexorable pursuit of his Holy Grail.

In the midst of panic in the night, even as the rubble was being searched for bodies, Benford began his inspirationally engendered program, critically examining the superstructure of nurses as they streamed back from the club. After a while, having spied no likely match, he subsided into dour disquiet etude; but finding no further signs of human carnage, he could also allay his fears that the owner of the bra might have been blown into another dimension.

The nurses were lucky, except for the one lightly wounded. And she was reputed to be a lesbian and thus of no interest to Benford. "And ugly besides," he said. And of most significance,

way too small, everyone agreed after careful
assessment.

The following day, after mulling over options,
Benford reduced his search to just the residents of
that one hooch. It was unlikely that someone had
been visiting and had shed her support garment
while there. Benford, being enlisted, thus poorly
paid and therefore of little interest to any of the
nurses as might otherwise choose to be benevolent,
had little success in approaching them directly. For
one thing, he was unknown to them. And, as I
said—and even nurses can be reasonably
perceptive—Benford was strange.

Several days went by with only stares and
curt replies the limits of information he was able to
gather about the Quonset huts residents. Over the
next few weeks, Benford spent all his free time
hanging about the Air Force compound, spying on
the nurses as they came and went. He was run off
a couple of times by the APs, and told to bugger off
once by a nurse Major with a large pistol which,
though not issue, Benford was convinced would
surely do him irreparable damage. He even
enlisted the help of other similarly fascinated, but
less vigorously addicted, fetishists, until he was put
on official notice by the Commander of the Air Force
medical facility.

But he couldn't stay away. The fascination
was too much. The bra, and its presumptive cargo,
had reached the point of idolatry for him, icon for a
soft, creamy, bouncy heaven of quasi-religious
significance. He'd been drawn to expansive
bosoms since puberty, he said, and hadn't the
maturity or mental capacity to overcome this
shallow compulsion—along with millions of other
GIs, of course. But Benford was committed!

The bra was stapled to the wall locker by his
bunk with a large, neat, hand-lettered sign stapled
above which read: "Who's Tits Fits?"

It was useless for me to point out to him the grammatically incorrect pronoun usage. And lack of subject and verb agreement. The meaning was clear. And like the Prince's minions, fanning out over the countryside, looking for the foot that fit the crystal slipper, every breast-bedazzled clown in CRAZY CAT was on DefCon IV to resolve the mystery.

Cathy the Captain, tight-mouthed as she is, let me in on the joke after more than a week had gone by. And it was a joke by this time. Every time Benford showed up on the Air Force side of Cam Ranh Bay, there were ill-concealed snickers and cruel teasing. But Benford had prey in his sights, or at least in his imaginings, and he was not to be dissuaded.

"Brenner, you know Barbara? The little brunette who assists in surgery?" Cathy asked.

"Sure." Not one of us it was who didn't know every nurse's name on the Air Force side, as well as at the Army's South Beach recovery hospital on the ass-end of the peninsula.

She sniggered. "The bra is hers." By this time, the entire population spoke of the '*bra*' in italics. There was no other *bra*. Anywhere. It had gained fame across the warp zone. "It was hanging on her bunk when the mortars came."

"Can't be. Cannot be! Barbara's a little skinny shit. Only about five-one, she can't weigh more than a hundred pounds. A chest flat as Kansas," I protested reasonably.

Cathy sniggered again. "Exactly. The *bra* was a joke gift to her from the girls. She's always—when under the influence of evil drink—spouting off about how she's going home through San Fran and getting herself some implants. Says she's going to 'beef up' like that Carol Doda who works in the Go Go Club on North Beach, the one who was in *Playboy*."

Whatever Barbara's intent, whatever the attitude of anyone else involved, the secret was kept. Nobody ever told Benford. Last I heard—I was already in the Delta when he came up for rotation—was that he extended his tour at Cam Ranh. Far as I know, he's still there, seeking the breasts that can fill out that magnificent piece of lingerie. Even the North Vietnamese's best efforts in Re-education Camp, after the war, could not have denied Benford his vision.

The Night Was Lit Up Like
a Football Field
By James 'Barney' Reynolds

About twelve years ago I was looking up chopper units on the Internet, and happened upon the 92nd Stallions and Sidekicks website. Retired WO Pilot Jim Koch, deployed to Viet Nam in 1967, from Ft. Carson, CO, is the webmaster. He had written a story about a night in Viet Nam that I quickly recognized. I was there the same night and the same place. Reading the story I realized WO Pilot Jim Koch was in the chopper that lit my detachment up like a football field that scary night. I didn't know if he was going to cost us or save us.

This happened on the night of January 31, 1968:

My detachment was Detachment 2, 330th RRC on Highway 1. We were two miles south of Dong Ba Thin, 30 miles south of Nha Trang, and across the bay from Cam Ranh Bay Air Strip in the year 1968. We were a 16 man DF Detachment.

At night we would have one guy pull guard duty in the tower that we built over the communication hut. Some nights some members of the Korean White Horse Division, which was located five miles south of us on Highway 1, would come out and go on ambush patrol west of our perimeter. They would also fire artillery randomly at different times.

This particular night we were in our hutches when we were startled by a loud, 'BOOM, BOOM, and BOOM'! Someone made the statement that the Koreans were really close! My buddy, Slim, the generator mechanic, and I walked outside to see what was going on. We saw a flash about 100 meters off our fence in the tree line, then we heard a SWOOSCH right over our heads, and saw the Cam Ranh fuel bladders behind us across the bay go up in flames! Those flashes sure weren't the Koreans firing!

Slim and I climbed to the top of our bunker and got on the net radio with Dong Ba Thin and TOUGH SWING 28. All the stations were reporting in and being hit. We could see where the flashes were located. All we had for protection were M-14s, an M-60, and claymores in front of the bunkers. Since the shells were flying above us and not at us, we decided to lay low. We called for the Reactionary Team, but we got the word, 'NO LUCK'. The 92nd had lost choppers due to a Sapper Team destroying five of their choppers. The Air Force runway was closed due to fires. The Koreans were taking hits on their line, also. We soon figured out, NO HELP was coming!

At that point everything became quiet. Slim and I lay frozen on top of the bunker, just listening, and watching. We thought we were hidden well until that chopper flew over and dropped flares that lit us up like the football field at the Super Bowl! I hugged the top of those sand bags. Man, did we try to look like a sandbag! I could see every goose bump Slim had on his arms! We stayed until daybreak and by then everything was quiet.

After daybreak we loaded up in the old deuce-and-a-half and headed up the road to Dong Ba Thin to eat breakfast. I wove the truck through the two barricades that were set up on the road. The barricades were being manned by GIs with rifles, claymores, and M-60s. We turned into the company area.

When we arrived at the mess hall area, everyone there were still in full field gear, laying low in the ditches. They informed us that they were still under 'RED ALERT'! They were still searching for members of the Sapper Team inside their compound. We just turned around and went back to our detachment, hungry, but feeling grateful for surviving another night at the LBJ Ranch!

I had the privilege of meeting WO Jim Koch at the Old Spooks and Spies Reunion in Colorado Springs in 2013. We had breakfast together and he met several veterans from our group. WO Jim Koch gave me approval to use his story in support of my story of the night of January 31, 1968. Below is his account of his experience:

January 31, 1968. Bob Meacham (My AC) and I - Stallion 505 flew 3.3 hours supporting MACV near Phan Rang. We carried mail, food, supplies, and did a few recons along the coast. Kind of a short day, but the unit was having a company party later and we were looking forward to it.

Wouldn't you know, upon arrival at Dong Ba Thin, the Operations Officer said that we had flare ship standby that night along with two Sidekick crews - UH-1C Gunships. We parked our ship in the 'hot spot' and we loaded up 40 Mark V Flares. There was no beer for the crew tonight, but the steaks were good! Shortly after 11:00 PM we hit the sack in the alert shack.

A tremendous explosion rocked the alert shack shortly after midnight! The alert siren went off and we heard numerous other explosions as we ran to the ship. Bob was in such a hurry that he ran out in his underwear and boots. He managed to run into the CO who told him to go back and finish dressing. While he did that, I started the aircraft and then we took off with the Sidekicks in under two minutes. I saw that several Hueys were burning on the north side of our airfield. To the east, at Cam Ranh Bay, the entire fuel storage area appeared to be in flames. We climbed to 2,500 feet and began to drop flares over Dong Ba Thin. I dialed in the Armed Forces Network (AFN) radio and there were reports that bases were being attacked all over the country.

With the light from our flares, the Sidekicks looked for targets of opportunity. Sadly, none were found. There were no incoming mortar rounds, rockets, small arms fire, or ground attack. At least five Hueys were burning in their revetments. After 2.8 hours of flight, we were ordered to land, refuel, reload with flares, and standby. The rest of the night was quiet.

AFTERMATH - January 31, 1968

Daylight had returned and what we saw was frightening. Three of our aircraft had been destroyed and one had been damaged by satchel charges which had been tossed in them by sappers. The 608th TC Co. had lost two. The VC had successfully penetrated our perimeter which was manned by Korean soldiers. Somehow they had also completely evaded capture. We would never again feel safe! Reports were coming in that many bases and cities have been over-run, including Nha Trang, Ban Me Thuot, Qui Nhon, Da Nang, and Hue. The first battle of the war, in Da Lat, was going on. The Sidekicks were going to Da Lat! That meant, among many other things, we would no longer get to R&R there.

My crew and I flew 9.1 hours that day. We saw things that we didn't think could or would happen. Numerous helicopters weredestroyed or damaged, many bases were attacked or over-run, and there were too many dead and wounded. The New War had begun!

Chinese Laundry:
Special on Greenbacks
By Robert Flanagan

January, 1965, saw the emergence of an R&R program to include ASA personnel, the 3RRU and 7RRU, in Viet Nam. I'll never know what tortured logic went into my selection as an early candidate for this treat—I was married, had a wife and three kids back in CONUS; carnal desires were off the menu; so was the financial wherewithal to enable such, should I have been interested, and I'd no money —but there I was, handed a set of orders entitling me to board an Air Force C-54 at Tan Son Nhut terminal on the morning of January 4 to proceed to Hong Kong for a period of six days, inclusive of travel time. I would be accompanied by one other 3RRU man, SP-5 Roger Cook, an analyst whom I knew well.

I was at the 10-month mark in my tour, looking forward only to a Freedom Bird home; before it was known as a Freedom Bird. As a husband and father of three, I had no money to spend and could only imagine the Crown Colony's exorbitant expense. So, initially, I declined. Then, a buddy, Jerry Lenihan, a soldier-for-no-one-ever-knew-what-reason, because he came from money, insisted I go and offered to front me the money. I already knew my subsequent assignment would be language school in Monterey; Jerry had come to 'Nam from 2nd USASA Field Station, Two Rock Ranch, Petaluma, Calif., little more than 100 miles north of Defense Language Institute. He had a wife and child there and, with political pull, would return there. He insisted we'd get together in California and I could then pay him back. Not believing in kicking a gift horse in the mouth, I told the first shirt that I was a go. He said he'd never doubted it.

We'd only been notified of our gift two days before departure, but on the evening of the 3rd, the night before we went on R&R, Roger insisted we go downtown for a meal at his favorite Indian Restaurant, the Punjab, or something equally Kiplingesque. I enjoyed Indian, though I could seldom afford to indulge my tastes. But with Roger assuring me he was good friends with the restaurateur and winking, said he would get us a good rate, I acquiesced. Taxi down to Tu Do, fine meal, couple of beers, and just before departure, when I came out of the latrine, I noted Roger and the owner in close conversation over in a corner.

When we left, I asked him about it. "Nothing. Just bargaining the bill."

"So, what do I owe you," I had to ask, as he'd picked up the tab and tip.

"Nothing. He didn't charge us. I'm doing him a favor." It was only then, in the back seat of the Renault taxi, that I noticed a brown paper-wrapped parcel, about 6 inches by 2-1/2 inches by 3 inches. Very compact. Sealed with tape. Unmarked. I stared at the package, looked up at Roger who wouldn't acknowledge my unspoken question, then back at the brown parcel.

Nothing more was said that evening; we lived in separate hooches on Davis Station, and we'd no chance for further discussion. When we got together in the mess hall the next morning, finishing coffee before grabbing our bags for the CQ runner taxi to the terminal, I asked him, "Rog, this is not something ..." I nodded into space; he knew what I alluded to, though it was nowhere in sight, no doubt packed securely in the most remote corner of his bag "... that's going to come back to bite us in the ass, is it?"

"What? I've no idea what you're talking about. I'm delivering a Vietnamese-made doll to the daughter of my friend's brother who lives in Hong Kong."

It was the last word he would say on the subject. And I thought, OK, give him the benefit. The Indian had bought us dinner last night. About the right size for a doll. Small doll. I can believe this, whether I did or not.

Two days before, at the start of the R&R program, two other 3RRU personnel had gone on R&R to Bangkok. Those were the only two choices, and then, not really choices. You were assigned to a destination arbitrarily, and if you would rather visit the other, you had to find someone on the R&R roster willing to trade whose work schedule permitted, who had the funds, which - Hong Kong was fine by me.

At this time, at the inception of the program, Customs didn't come into it. No one, not an MP or an AP or a civilian Vietnamese factotum, asked to check our bags nor asked us questions. When we boarded the plane, the Air Force pilot announced on the PA that if any of us were carrying weapons or live ammo or any kind of ordnance, to make ourselves known and a flight attendant would visit with an unmarked box into which could be deposited all the above, receipts available. I don't recall any takers. The brown parcel rested undisturbed in Rog's bag.

Following a chancy landing at Kai Tak airport, with its cross-wind strip protruding into the bay, we deplaned and were bused to the Park Hotel on Kowloon side, the headquarters for all R&R activities. After check in there, and briefing by both Army and Air Force NCOs, we were free to choose any other hotel, but were admonished to keep in contact with the Park where any recalls-to-duty or announcements would be posted. Roger and I decided the Park fit our needs, which were minimal; it was well-located on the mainland side, the R&R rates were good, and we just didn't want to spend precious time seeking out other lodgings.

We had adjacent rooms, and by the time I'd hung up my extensive wardrobe—the one spare pair of trousers and two shirts, along with the short-sleeve khaki uniform I'd traveled in—Roger was banging on my door. "Let's go."

"Where?"

"Gotta make a delivery." He was carrying a plastic shopping bag with a suspiciously shaped something suspended in it. I'm OK with that; I thought; the sooner we divest ourselves of this 2-lb. pack of cocaine or Mescal or marijuana, all the better. I walked some yards away from Roger when we went out onto the sidewalk.

I thought we'd grab one of the gazillion taxis that flooded the busy streets, but my mule said he had directions and it was only a couple of blocks to our destination. We walked up the street to Kennedy Road, not JFK; the Brit overlords had a passel of Kennedys they thought worthy of naming colonial streets and roads after. The farther we walked, the longer we were exposed on the street, the larger in my eyes Roger's plastic bag got. I thought it had the multi-leaved logo of the Cannabis cult printed on it, and wondered at the absence of a Kung-Fu qualified Chinese narc who should, by all rights, be tracking us.

Roger was checking street numbers on shop fronts, and soon we came to the one he sought, a neat-looking enterprise specializing in stationery, writing materials, quality fountain pens and cheap ballpoint pens, a few touristy key chains and an endless display of postcards of everywhere but Hong Kong. A heavy steel security flex gate was locked in place across the entire front, door and shop windows. I threw up my hands: "Is this the place. Great! It's closed."

"Not to worry, Kemo Sabe. Pendar said to knock; we'd be expected." I suddenly realized there were depths to this operation I was totally unaware of. Roger rapped smartly on the door frame.

A shade inside the door was pulled aside and a wary eye peered out. The shade dropped; the door instantly unlocked, cast open.

"Roger! Are you Roger?" the Indian man asked with a huge smile. When they'd shaken hands, and I was introduced, the Indian said, "Danny Ranesh, So pleased to meet you both."

I felt conspicuous, standing there with the door open, obviously now concluding part of a major illicit enterprise, and our contact seemed not to recognize the dangers. I inched my way into the darkened shop, willing the other two to accompany me. Danny, perhaps sensing my discomfort, shut the door, leaving the shade in place.

He said, "Are you settled all right?" His English was perfect Brit colloquial. "Does the Army put you in a place you like? If not, we can find you alternate quarters. It would be my pleasure."

"No, we're fine at the Park. Convenient," Roger said. "And we'll only be here five nights."

As he spoke, he moved the plastic bag in front of himself, proffering it to Danny. The Indian reached and took it casually, placed it on the empty sales counter and stuck his hand into the bag, pulling forth the iniquitous brown parcel. After a quick glance, he reached for a sharp-bladed, small letter opener beneath the counter and said, as he began dismembering the taped paper covering, "I thank you so very much, Roger. And you, Bob, for this act of kindness." I began suddenly to credit the doll fantasy.

"This is such a great thing you do for our families, Pendar's and mine." In almost concurrent explanation with the casting aside of the wrapping materials, Danny both destroyed the doll myth but relieved my drug-washed fears, while leaving a slight discomfort yet.

"The currency laws of the Vietnamese government," he said, "make it impossible for merchants like us to exchange, even within our families, mutually independent earnings."

I was later to learn that Vietnamese law forbade the export of any currencies not of Vietnamese origin. And no one would even try to export Viet currency, as it was practically worthless outside its own country. American dollars, which were heavily, though illegally, traded on the open market in-country Viet Nam, were especially valuable; and were acquired by the ton by book sellers, restaurant and store owners, whores, bar tenders, house girls ... almost anyone but legitimate merchants in Viet Nam. Currency crimes brought heavy penalties, including on occasion execution, but more often imprisonment, loss of homes and businesses, and deportation for non-natives.

Suffice it that the government of Viet Nam lost a nice little bit of lucre when my friend did his restaurateur friend a favor. Looking askance at the stack of currency, I estimated between $8,000 and $13,000, depending upon the inclusive mix of $20s, $50s and $100s.

Danny was open about displaying the illicit currency to us without comment, and he locked it into a wall safe behind a curtain in his shop. He picked a bottle of Johnny Walker red label from a desk, put it into a brown paper bag, and locking the shop door casually behind him, led us up the street to his club, an unprepossessing little gathering place with five or six tables upstairs above a billiards parlor. The food was entirely Indian cuisine—fine by me—but was served without benefit of a menu. Danny ordered dinner for three,

and we ate what was brought. The strangest thing I
recall about the occasion was that no liquid of any
kind was offered by the service staff or, I gather, the
establishment. I'd expected perhaps an IPA or
some European beer, but not so. Not even tea.
Nor water. It was the anticipation of our host that
provided the bottle of Johnny Walker red, though an
unlikely quencher for some of the spicy Indian
dishes. But the meal made up for my aforesaid
squeamishness, and Roger and I went for several
more free meals at the Punjab in Saigon before I
was forced out of country to return to the land of
Pizza and grits.

Chapter 4
Time and Place is Everything

A Cautious Trifecta
By Robert Flanagan

Present in the mixed Irish-Cajun genes of my existence is ample opportunity for the belief in, and employment of, various spells, curses, heavenly divinations and suspicions. Yet I remain amazingly superstition-free, tending to fall back on hard reality in instances where less confirmed skeptics would fall prey to the dark side of our imaginings. And yet there have been times when that blithe, offhand manner has been threatened, when I have felt the insistent stirrings of *something out there!*

This lapse of equanimity most prominently presented itself in a series of events that befell me in Viet Nam. One close call, or if not exactly that, then a challenge of fate, can be passed off as normal circumstance. Two, a more unusual escape from the dark side. But when a third portent of destruction is narrowly and completely unknowingly averted, it's time to re-group.

In the fall of 1964, I had been promoted out of an A-Slash job at White Birch, and as a new Staff Sergeant transferred to become the NCOIC of the ARDF Section of the 3rd RRU. I quickly fell among thieves, linking up with a couple of other Staff Sergeants who were also section heads. John Moore was NCOIC of TA, and J.T. 'Jerry' Lenihan was NCOIC of Management. We became a grouping, essentially a threesome of sometime-malcontents not always in complete accord with the holy writ of 3RRU diktat. And, too, within the Operations hierarchy of 3RRU was a new Ops Officer, 2LT Gary Chladek.

LTC Thomas S. Owen, 3RRU Commander, had issued an edict: only officers could sign for a vehicle which was most often a Jeep; but no Officer could drive a vehicle. As Ops Officer, LT Chladek was issued a Jeep, but had to solicit a driver from among we enlisted swine. Our trio jumped at the chance to get partial ownership in transportation, and it became common for one of us to drive the lieutenant, mostly on personal forays. Along with this perk, we developed the habit of going for lunch at the airport's small civilian terminal café, labeled the Viet Nam Air Restaurant, just up the flight line from our headquarters in the hangar. The four of us became a fixture for lunch. I may be a few piastres off, but I recall a small steak, fries and weeds (Moore's term for salad), and a *Ba Muoe Ba* was about $1.25. What's not to like?

On November 19, I had scheduled myself for an afternoon mission flight; takeoff time was about 1230 hours, and I couldn't afford the time for lunch with my cohorts. I had caught a ride back to Davis Station, chowed down on some of Walt Gentry's renowned cuisine, and was walking back up the flight line when I heard an explosion and watched a dark cloud of smoke and debris rise up before me, some quarter mile away. I knew it had to be VC inflicted, but I hadn't a clue as to target.

When I landed and de-planed after my mission, my clerk informed me that Chladek, Moore and Lenihan had been blown up in a mined explosion in the restaurant, right at 1200 hours. A block of *plastique* of some estimated 40-45 lbs. (some rumors had it as a captured American anti-tank mine) had been planted above the ceiling in the center of the restaurant, and either timed to explode at noon when the restaurant would be at max, or was remotely fired when such condition was deemed most advantageous. In addition to the three ASA men, 11 Air Force personnel and three children were injured; one Airman, a Lieutenant Colonel outside the building on a bus, was severely wounded. No deaths were ever confirmed, though I cannot attest to what occurred downstream, out of my region of contact.

All three 3RRU men were alive, but suffering from multiple lacerations from blasting chunks of concrete and perhaps shrapnel and both Lenihan's eardrums were ruptured. Neither of the other two could hear in the immediate aftermath. I visited Lenihan that evening in the Navy Hospital in Cho Lon; he was swathed in white bandages and his arms looked like he'd been tossed whole into a pit with a dozen rabid Rottweilers. As I remember, John Moore had a severe gash on top of his head, and his arms, too, were chewed up; he also may have had a ruptured eardrum. So, too, was the lieutenant. They'd all three sat with their arms resting on the table awaiting service; and it was determined that they were seated directly below the blast device. The reason they weren't more seriously injured was likely the cone-shaped blast that spared them the critical center of the explosion.

John Moore was shipped out to the PI for treatment; he returned in about three weeks, if I remember correctly. J.T. Lenihan was sent to Letterman Army Hospital in San Francisco, and from there, to Two Rock Ranch where he owned a home in Petaluma and had a wife and child waiting. LT Chladek was also shipped out and I lost track of him; I later served with him again as Major Chladek at VHFS, 1973-74. He remained for a career.

How many times I've played over that scenario: that if I had been there with that tight-knit group, as by rights I might have been except for the peripatetic scheduling of myself on a flight—what my results might have been. Maybe just a Purple Heart, cheap at twice the price, if I'd sustained no serious damage; or it might have been the whole enchilada. But it was just a happenstance. Nothing intuitive; nothing perceived as exceptionally fortuitous. That was then!

In early January, 1965, two of my ARDF ops went down to the casino, off Tu Do Street, and drank an indecorous quantity of *Ba Muoi Ba*. Dan Bonfield and one other Op—I can't remember who—returned to TSN, took a cyclo across the base toward Davis Station, but got off the cart 100 yards-or-so before reaching Davis Station gate. In their cups, they'd decided to make an attack on the ARVN Ranger Training facility across the road from 3RRU. Bonfield and cohort crept along the dark road, and when they got within 30 or 40 yards of the compound, leaped upright and charged, screaming in faux movie-script Oriental macho lingo: "Die Amelican dogs. You die tonight. We your worst nightmare, round-eyed dogs," and so on.

On the corner of the ARVN compound at the intersection of the roadway was a sandbagged machinegun emplacement. Normally not manned, it had a three-inch steel pipe imbedded in the ground for quick mounting of a machinegun at times of crisis. Despite inebriation to an advanced degree, both charging warriors had enough awareness, at some point, to realize they might have opened the wrong can. No ARVN guard was present, but there were lights within the compound, and just across the road was stationed an armed 3RRU guard. At the sandbagged pit, both assault specialists found it prudent to take cover; they dived over the sandbags into the gun pit. Bonfield crashed, knee-first, into the steel pipe; the other warrior smacked himself unconscious against the sandbags. The Third's gate guard, by the mail shack, strolled across the roadway to ferret out just what the hell he'd been witness to, and it was he who called the CQ and the ambulance from the 32nd Air Force Dispensary to come for their charges.

So, that's all prologue. On the 18th of that month, January, Bonfield limped into the scheduling arena in the Hangar and said, "Sarge, you gotta get me on some flights. I only got x number of flights in (I think it was, at that time, two), and I could lose my Hostile Fire Pay this month."

Flying crews drew $65 a month Hostile Fire Pay then, on the presumption of being under fire in flight; but everyone drawing the extra pay had to fly at least six days in a month to qualify. If you didn't fly six, the $65 was forfeit.

Bonfield could barely walk; his knee was still about the size of a volley ball. I told him no. He pleaded. I threatened. He continued to plead. Up to my knickers in scheduling ARDF efforts—3RRU and the ARVN complement—I relented. "OK, Dan. You can take my flight scheduled for an AM mission tomorrow. But you gotta walk to the bird. No being driven out with your gear and all; you gotta be able to walk."

What the hey? We'd never lost a plane. There was no reason to think four hours sitting in the back seat of a Beaver would put too much strain on his ill-gotten wounds.

On January 19, WO1 Ray McNew and SP-4 Dan Bonfield broke ground in an RU-6 "Beaver" about 0800 and flew into the ether to gather dits and bearings. A couple of hours later, working at my tasking; I was called to the phone. It was the Air Force Air-Sea Rescue Unit, calling to ask if we had an aircraft, tail number one-five-one, flying anywhere in Three Corps. I knew 151 was McNew's bird, but I couldn't answer the lieutenant's questions as to where the aircraft would have been flying. I looked at my watch, checked Bonfield's tasking sheet, and knew where he *should have been* about the time the Air Force had monitored a MAYDAY call from 151, "... going down on a plantation strip near Tay Ninh." As we flew classified missions, we didn't file flight plans, so Air Force had no record of where the bird's mission was.

I called Captain Carlisle, Air Section CO, reported the situation, and he instructed me, based on knowledge of the sked, to "find an unoccupied pilot, grab a bird, and go help in the hunt." CW3 Martin was available, and within ten minutes, we were airborne in an RU-8. I gave the pilot the location I thought most likely—a decent sod airstrip on a Michelin plantation near Tay Ninh—and we plowed on across III Corps with great hopes. We called another ARDF mission bird, which had been working down around Vung Tau and was on their way back in after mission, diverting them also to the hunt. But as Nui Ba Din grew large in our windscreen, we heard suddenly a broadcast from ASR: "All aircraft in the vicinity of Tay Ninh or the Parrot's Beak depart the area for the conduct of aircraft and personnel recovery."

We banked and held in a lazy circle until we saw helicopters rising from the distant tree line, and then headed for the barn. After landing, taxiing, and shut down, I returned to the Hangar where the S-2, a young 2LT whose name is lost in the mists, soon sought me out. "Sarge, we've got to see the pilot and operator."

If I thought concern for the well-being of the two soldiers was guiding his interest, I was soon disabused of that frivolous notion; I realized he was concerned for the COMUS pad, a highly classified, one-time encryption pad which, in a time without secure communications equipment, allowed an airborne operator to pass a critical message to the ground in a secure fashion, though slow and ungainly. On each flight, the operator signed out a COMUS pad and it was his responsibility to return it to the safe in Operations after mission.

We found Mr. McNew sitting on a hospital stool in the 32nd being bound with Ace bandaging, and Bonfield unconscious on a stretcher. We got from the pilot the details of the mechanical failure which led to a hotly contested powerless landing on a sod strip; their rescue by Ruff Puffs, and extraction by a US Huey. McNew, as I remember, had some cracked ribs, maybe his collarbone, and terrible bruising across his mid-section from smashing into the yoke when the bird struck the two rows of rubber trees. Bonfield had broken bones, in addition to the magnificently swollen knee from the 'attack,' and oddly, a day or so later, was found to have a bullet lodged under a shoulder blade, a slight anomaly missed in the early triage. But it proved to be a false concern on the S-2's part; Bonfield, knowing what a pain the COMUS pad was to work with, and convinced that nobody on the ground paid attention to most DF fixes anyhow, chose never to carry a pad on the bird. We found the pad where it lived, in the S-2 safe.

But again, if I had been where I should by all rights have been—flying my own scheduled mission with McNew instead of being suckered into playing to Bonfield's entreaties—would I have been just a slight space over and caught that round in the heart or head, or had an R-390 punch through my chest? Again, it was just a bit of weirdness. But it was the second one: in both cases I wasn't where custom would have placed me in any undisturbed sequence of events.

Then a period of respite: the third race in this beguiling trifecta came during my second tour in 'Nam.

Late summer, 1969, Cam Ranh Bay: Flying Controller on CRAZY CAT missions, I was on the manifest for a flight on the following day. Our usual flights were 13-15 hours long, including at least eight hours on target, and a lot of logistical time prepping, getting there and back, so when one was scheduled to fly, it was best to have a night's sleep under your belt. There was a sort of informal rule that if scheduled, you were supposed to be in your bunk before 0200. This was not rigidly enforced, but it was to one's advantage to play straight; the stress of missions that long in aircraft that were questionable as to airworthiness tended to drain one's resources pretty quickly.

This night, before my scheduled mission, I had fallen prey to an insidious band of miscreants, a poker game with pilots—a nefarious lot. At Cam Ranh, we billeted in standard barracks of two stories, with individual suites, three small spaces in one-room area shared by two officers. The poker game was on-going in the building adjacent to my billet space, on the far side away from my room. It was a tense game; both the beer and cash were flowing.

Along about oh-dark-thirty-seven, we took a quartet of 122mm Soviet rockets which rained down fruitlessly on our area. Their targets were obviously the highly valued aircraft—if not our lumbering old P-2Vs, then the Air Force's fast movers next door— but Charlie was not especially adept at rocket barrage warfare. I think our security force, the Korean White Horse Division, might have had something to do with his hurried firing protocols. In any case, there were no aircraft hit, no casualties, just a lot of huff and puff and running for the bunkers and sweating out the Korean's response 155mm fire, far into the night.

When the All Clear was given, and we emerged from the bunkers, the poker game was a lost cause. I had my hand of cards in my pocket, along with my chips; but not all players were as prescient. Cards, chips, beer cans and bottles were scattered over the room; and all inspiration in the game was gone. We wandered away to our rooms.

When I opened the door and turned on the light, I found myself in a snow storm. White flakes whirled in a furious bizarre spiral about my head, settling on everything without melting. It took a few disorganized seconds to realize it was not snow, but feathers. From the center space between our two rooms, I looked in at my area. In a severe shift in the plane of normalcy, it took me a while to fully assess circumstances. My pillow was a flat casing with a few fluttering feathers about it, lying atop my blanket-covered mattress, but the blanket was in two pieces. The bunk was still nicely made, but someone with a scalpel had slit my blanket up the center, from the foot of the bed to the pillow, leaving only a few inches where the blanket was still in one piece.

When a friend dropping by to ask me something about the next day's flight noticed a small puncture in my outside wall, it was an easy, though disquieting, thing to track the course of the piece of shrapnel. The 122 had landed between our hooch and the one next door housing Navy MARKET TIME pilots. It blew a pit in the ground big enough to dip sheep, but seemed otherwise not to have damaged anything. The Navy billet was unoccupied at the time, and few people were in the 1RRC billets, me included.

A piece of shrapnel had torn through my wall, ripped up the center of my bed, parting the blanket and exploding my pillow before burying itself in a 2x4 wall stud. That piece of Urals cast iron, ragged and ugly, is in a small framed case in my memorabilia, a deft but uneasy reminder of this, the third instance of my Viet Nam service in which, had I been where I should have by all rights been, somebody would have been inventorying my gear the following day, and my upcoming DEROS two weeks later would have been pushed up in cavalier fashion.

But where do you draw the line? Fate? Luck? Karma? Call it what you will, but given those odds again, I'd not advance a single piastre against them.

An Answered Prayer
By Juan C. Mendoza

It's just another day at the office in Phu Bai. We
started copying one of the Viet Cong units and
taking directional shots, just as we always did. Then
someone intercepts a message that gets everyone
excited. The word gets out – sleep with your boots
on. Sleep with your boots on was a message that
let everyone know that a rocket attack was possible.

It was June, 1969, and I am looking forward
to heading home. I was scheduled to leave Viet
Nam around the third week of June when the word
gets out sleep with your boots on. I pray, not now, I
am going home. Please lord don't let this happen. I
start getting that uneasy feeling that something is
going to happen to me, that I might not make it back
home.

We're sleeping in the trailer when all of a
sudden – BOOM, we hit the floor of our room. We
wait and then BOOM, BOOM. We get up, get our
gear and run out the door to the bunkers. Rocket
attacks in Phu Bai were rare but they did happen.
And, there were usually only three hits. We wait in
the bunkers and trench line until the All Clear is
given.

As I am walking into my room I step on
something hard. I look down and pick it up. It's a
piece of shrapnel. The next day we were surveying
the damage and see that one rocket hit just in front
of the club house and destroyed the front wall. I
don't remember exactly where the others hit, but
shrapnel from a rocket tore through the trailer walls.
When we got back to the room and looked at the
hole on the wall we tried to project the path. That's
when we realized that if that piece of shrapnel had a

little more force to it, it would have hit me as I lay in bed.

I thanked God for hearing my prayers earlier and allowed me to go home safe and sound.

Thinking of how close I came to dying has haunted me for 46 years. And, every time I think of it, I thank God for answering my prayers.

This Wasn't Supposed
to Happen In the ASA
By Ed Johnson

During the mid to late 60's and early 70's, a majority of the U.S. Army recruiters didn't always tell the whole truth about the ASA in Viet Nam. Their usual line was that if you went to Viet Nam, you would be in a rear area, living and working in an air conditioned environment, etc. Numerous recruiters would neglect to tell potential enlistees about Direct Support Units supporting the divisions, separate brigades, and a Special Forces Group. For the record, I would like to state that my recruiter informed me about DSUs and their mission.

In April, 1969, Operational elements of the 3rd Platoon 265th RRC deployed with the 3rd Brigade, 101st Airborne Division to conduct combat operations in the A Shau Valley. The 3rd Platoon supported the 3rd Brigade throughout the Spring and Summer of 1969, including the battle for Dong Ap Bia, aka Hamburger Hill.

The 3rd Platoon's bunker was located just outside the 3rd Brigade Tactical Operations Center on the west side of Fire Support Base Berchtesgaden, which we just called BG. We had a CONEX that was set up for transport by CH 47. Inside we built counters along the three sides for our intercept operators. There were two field desks by the door, one for the Platoon Leader and one for the Traffic Analysts.

A trench had been dug or blown, and the CONEX was dropped in by a CH 47. We then placed 8" x 8" x 10' or 12' beams across the trench. Over the structure we laid pierced steel planking, a layer of sandbags, a rubber membrane, and several more layers of sandbags. It was approximately 8' W x 6' H x 20' L. The entrance was on the east end of our bunker, just a few feet from the west entrance to the 3rd Brigade TOC, and down a steep set of steps dug into the ground. Along the sides we had built bunk beds for 4 from scrap 2 x 4s and 105 mm ammo boxes. We basically hot bunked between guard and Special Intelligence - SI duties. We didn't have enough bunks for each person. On average there were probably ten to twelve people at BG from the 3/265 RRC.

Fortunately, SSG Bruce Heidemann, the 3rd Platoon Sergeant had us build a blast wall out of sandbags at the base of the steps. It was annoying when we were entering or exiting the bunker because the blast wall took up space at the entrance.

Around midnight on June 13, 1969, I came in from several hours of guard duty and changed into dry clothes and crawled into the bottom bunk on the right side of the bunker. There was only minimal clearance between the bunks. I was blasted out of my sleep when a satchel charge exploded. The satchel charge landed in front of the blast wall, which deflected most of the force out of the entrance to the bunker. The blast bounced the left side of my head against one of the bottom supports of the top bunk and me onto the ground. John Braun was also blown off the top bunk. At that point I had 'Excedrin Headache 101'. I remember laying on the ground seeing stars while John Braun was stepping all over me while he was putting on his flak vest, LBE, etc. In those days we slept with at least our boots and pants on.

The bunker was filled with dust so thick I couldn't see, but I could smell the cordite from the satchel charge. I believe it was Dale Renner that pulled me outside. For a short time I could not tell what time zone I was in. I was led to one of the defensive positions on the west side of BG. Somebody noticed I was bleeding from the back of my head and neck and applied a field dressing.

We had a communal shaving mirror that we kept near the entrance. The blast had shattered the mirror and embedded fragments of the mirror, dirt, and gravel from the side of the bunker into the back of my neck and head. I also had a lump on the left side of my head above the temple. Someone took me back inside the bunker after the perimeter was secured. At first light the MEDEVACs started coming in, and I was taken first to the 22nd Surgical Hospital in Phu Bai, then out to the USS Sanctuary.

COL. Conmy, the 3rd Brigade Commander, was also wounded in the attack and came by and visited with us while on the USS Sanctuary. He was one of the best commanders I ever served under, and he became the standard by which I judged my superiors in the future. I got to know him fairly well as a SP-5 because I used to take Special Intelligence information to the Brigade TOC. Only the S2, S2 NCO, S3, and Brigade Commander had SI clearances. We usually gave SI info to the S2. If the S2, S2 NCO, or S3 wasn't in, we would see COL. Conmy if the information was time critical. Also, while out on LZ's or operations, COL. Conmy would usually stop by and talk with us. He was always gentlemanly and professional when he spoke. He was never condescending or rude to the enlisted ranks. I read in the Screaming Eagle around 1992 or 1993 that he had died.

A week after the attack I returned to BG from the USS Sanctuary. One of the guys I knew from the Headquarters Company 3rd Brigade Communications Section said that he saw flames shoot out of our bunker from the reflected blast and thought we were all crispy critters.

Several days after coming back, we were again attacked by the NVA on June 22, 1969. One of our positions was a horseshoe shaped parapet made from sandbags about four feet high so we that could look down a ravine on the south side of BG just below the lower helipad. Phil Wade and I were on guard duty around 0200 - 0300 when BG started getting probed on the east side. Shortly afterwards, a flare ship was up and illumination rounds were being fired from several Fire Support Bases. There were several sappers wearing only loincloths below us. We fired short bursts at the sappers but don't believe that we hit them. There were other NVA outside the perimeter firing rifles and tossing grenades. Sometime later a flare went off, and Phil Wade saw a NVA throw a grenade toward us. Phil, who was to my right, pulled me close to the inside of the parapet. My right hand extended out over the parapet, and the right side of my face was partially exposed. The grenade landed 15-20 feet from us and exploded. At that time I didn't think I was wounded. We went back to being vigilant, and the situation calmed down as the NVA withdrew.

A little later I was feeling woozy and told Phil that I didn't feel good. He checked me over and said, "You're bleeding from your right hand."

He called for a medic. Lt. Pintozzi from the 326th Engineer Battalion came over with his medic, who then treated me. Lt. Pintozzi took me to the TOC for further medical attention and evacuation. The grenade fragments had hit me in the right hand, face, and shoulder. The vein was perforated near my right middle finger and after 35 years I still have eight or nine pieces of shrapnel in my right hand, face, and shoulder. Also, the tops of Phil Wade's and my camouflage helmet covers were shredded.

I was MEDEVACED out at first light to the 85th EVAC in Phu Bai and subsequently transferred to the 6th Convalescence Center (6th CC) in Cam Rahn Bay for a month after surgery on the vein in my hand. In early July, NVA sappers attacked the

6th CC. A satchel charge was thrown into the opposite end of the ward I was occupying. Fortunately, no one was injured in our ward, and I don't believe there were any casualties at the 6th CC as a result of the sapper attack.

I returned to BG in late July for a short time. Several days later, I joined a team on LZ Rendezvous until we left the A Shau in mid-August, 1969.

I believe that if SSG Heidemann hadn't had us build the blast wall, the concussion from the satchel charge in such a small, confined area would have killed most or all of us.   There would have been a dozen 265th RRC names on the list of cryptologists killed in Viet Nam on the Memorial at NSA.

Next It'll Be Rum
By Robert Flanagan

On November 19, 1964, I was scheduled to fly an afternoon ARDF mission, launch time 1230. As one of Sergeants Three—Ops section heads—I had been accustomed to making lunch most every working day with Staff Sergeants John Moore of T/A and J.T. Lenihan of Mgmt, along with 'the kid' (Lenihan's title for our Ops Officer, 1LT Chladek) at the small restaurant in the civilian terminal building on Tan Son Nhut, at the far end of the ramp from Davis Station. LT was included primarily because only officers had Jeeps—but officers couldn't drive. The Colonel's edict afforded transportation for whoever could exert enough influence on vehicle owners. That day, though, I was to miss that mid-day gathering; I'd hopped a ride back to Davis Station billet area to partake of Walt Gentry's cuisine, early chow in the NCO mess, because of imminent mission launch.

Returning to Flight Ops, walking up the flight line at twelve hundred hours, I felt the effect of a distant concussion; I sensed it was a telling explosion. Following the eruption, directly ahead I watched a cloud of dust and smoke and debris lift into the sky beyond the line of aircraft ranked before me. It looked to be somewhere in the civilian area of the airfield, and I heard, quickly, a chorus of sirens and indistinct PA panic.

No one in Flight Ops knew anything; they'd hardly noted the concussion, shielded as they were by buildings and by indifference: If it ain't here, it ain't important! I grabbed my flight helmet and mission bag, shouldered an M-14 and followed Captain Carlisle out to the flight line. I was flying with the boss today, both of us—he as Air Section OIC, I his NCOIC—pushing administrative duties into smaller and later corners, while daylight hours were mostly given over to flying. Since at that time visual orientation for fixes required daylight, Third Air still was essentially a daytime job—though sometimes it strained the boundaries on either end.

After four and a quarter hours flight time, we landed and were walking across the apron back to Flight Ops when we passed SP-5 Mike Campbell, an analyst from WHITE BIRCH who liked to hang out with the Air people. Campbell saluted the Captain, and asked me, "Hear about your buddies?"

I'd heard nothing but static and erratic VC comms for the past four plus hours, and said, "Now what? What buddies?"

"Your lunch crowd. They all got blown up, over to the terminal."

"What the hell are you talking about, Campbell?" I demanded, the Captain and I stopping short on the ramp.

"What's happened, soldier?" Captain Carlisle demanded, recognizing Campbell's game for what it was.

"Uh, yes, sir. Sergeant Flanagan's friends, Sergeant Moore and Sergeant Lenihan and the Lieutenant . . . they were havin' lunch in the terminal. Straight up noon, a bomb planted in the ceiling of the airport restaurant went off. Killed some people. Nobody seems to have the straight of it yet."

"What about our guys . . . any casualties?" I had to know.

"I hear they're all in the hospital. None of them got killed, I don't think. Busted up pretty bad. Some Air Force Lieutenant Colonel got killed, and an MP. A couple of Zip waiters, I think." He seemed uncomfortable now with his knowledge.

The Captain and I broke into a trot for 3RRU Headquarters.

* * *

Moore had been taken to the Air Force 32nd Dispensary on Tan Son Nhut. I found him propped up in bed, his head swathed in a crusty-looking bandage-helmet; bandages on his arms and shoulders and brown ointment thick on hands and face. One eye — the right, I think — was covered with a bandage. He was conscious but seemed to be drifting in and out of lucidity, likely from the sedatives given for pain ... or, it could just have been John's acerbic personality.

"Hey, hoss," he said dreamily, "to use the vernacular you're so fond of. How'd you find me here? By the way, where the hell is here? No one's said. Nobody talks to you in these places."

"The Thirty-second. Under the gentle ministrations of our comrades in harm, the Air Farce. And up your vernacular too, you wretched man. How they hanging, John?" Expression on my face must have been an ephemeral thing, shifting, as I felt myself responding to every change, every blink of his good eye, bouncing between anxiety and worry—a narrow choice.

"Thank Christ, they're still hanging." It wasn't one of his shining moments, erudition-wise: Moore's tendency toward sarcasm and irony was well established. I guess he couldn't get the old spleen up and running, or whatever in Oz it was that kept him humming along. Captain Carlisle came in behind me, leaned over and spoke to him. I couldn't understand then, don't remember now what he said; something solicitous and meaningless, no doubt. We understood by then John probably wasn't going to die and mess up flight skeds and the personnel roster.

"Do you know what happened?" I asked. At his request, I had to shout into a cupped-ear.

"We got our shit blown away. What else?"

"Can you remember any details, dip stick?" I felt a quick need for something I wanted to say, something funny, something brilliant, but lost it before it got to the lips.

"We'd just walked in. Sat down. The three of us, sitting there, jawing. Had a beer. All of us," and as he said it he glanced at the Captain to check his response—Chladek had been under an alcohol-watch edict from Carlisle following New Guy Indiscretions at the hands of evil NCOs testing the LT's limits right after he arrived— but Carlisle didn't appear to notice now.

"You'd skipped out, on us, you fink," Moore made the point to me. "Bomb must have been set for right at twelve noon. No warning." There was a momentary lull as he went off into la-la land. Then, "I don't remember hearing the blast; now I'm not hearing anything all that great. I guess it was directly over our table. I heard some air dale Colonel bought it. And some little indigenous people."

"Yeah, and an A.P. from Air Force Headquarters, I think," I reported what I'd heard. "Was it a bomb or what? A grenade?"

"No, they think bigger." He concentrated for a moment. "Ordnance guy in here earlier said they thought it was an anti-tank mine. Probably one of ours. Or maybe French, left over from ten, twelve years ago." Another pause. "The fact the building was built by the French, had steel rebar in the concrete, and is probably what kept us from checking out." Explanations and rationales; always in demand.

With a sudden spurt of energy, Moore asked, "What do you know about the other guys? I can't remember anything until I woke up in here, but one of the medics said they thought El Tee was brought here, too. How about Lenihan?" Where else could he be . . . if he was alive?

"Yeah," the Captain said, "Chladek's here, but he's unconscious still. They're prepping him for head surgery. He got a piece of shrapnel in the wrong place. You two here. Lenihan's in Cholon with the squids." Carlisle had found a source of info somewhere in the hospital. "But you guys were all lucky; most of the damage seems to be the tops of your heads, and forearms—sitting with your arms on the table, I guess. And you were lucky to be directly under the blast. Most of the shrapnel blew out in a cone shape. That's what got the other casualties. 'cept I think the Colonel was outside, sitting in a bus, awaiting transport.."

"Yeah," John mumbled, seemingly suddenly surprised to be speaking, "aren't we lucky!"

Chladek, it later became known, had emergency surgery that night and was shipped back to Letterman General Hospital, San Francisco, for further surgery and recuperation. And he did recover, but he never returned to the Third. Nine years later, I worked with Major Chladek at VHFS. SSG Moore was sent to the Philippines, worked over, and by the time the head wound — slight — and the arm lacerations — relatively unimportant — were healed, his hearing and eye had also improved so that he was returned to duty by February.

Carlisle was right. Lenihan had been taken to the Navy Hospital in Cholon, both eardrums ruptured, and he had an ugly, semi-severe head wound, as well as arm, shoulder and neck lacerations common among the three.

With the LT out of service, his Jeep was up for grabs; I grabbed it and went to visit Lenihan that evening, but he was sedated. I learned from a Navy nurse that he was scheduled to be shipped to Japan for surgery in a day or so, probably on to Letterman also, since he had homesteaded at Two Rock Ranch near Petaluma for years, and had family and a home there.

Two days later, curtailing the quiet, cool, early morning bliss of a Dawn Patrol, reluctantly disembarking to the inhospitable earth, I returned to Operations and read the message stuffed in my distribution box. The clerk had written: SSG Lenihan left word URGENT! : "Please come Cholon Hosp. today. Will ship Japan/CONUS tomorrow. Jerry." Repeat URGENT!

Jerry? Who the hell was Jerry?

I was stumped, stood frowning into the void, trying to match the name with someone I knew. First to mind was Jerry Baines back home, reputed to be a cousin of some derivation, though relationship was denied by family members. But that was ridiculous! Baines had evaded the pitiless grasp of the draft, fleeing at oh-dark-thirty on a north-bound train, and the last I had heard my erstwhile kin was living in a commune of Lithuanian lesbians in Calgary. And the clown only had three toes on each foot; the military wouldn't have taken him on a bet.

My mind searching, my eyes roamed over the chalkboard above the clerk's desk. Thou shall maintain thy airspeed, lest the ground rise up and smite thee mightily. Aviator humor; Losing its appeal with each succeeding flight into the beyond.

Jerry. Who the hell? I looked at the clerk's note again: 'Lenihan left word.' Jesus, Flanagan! shaking my head to clear it of fuzzy, high altitude clutter. Get one's head out of one's ass.

Lenihan! I stared at the yellow sheet, gears grinding exceedingly slowly. Blind spot! Of course, the message was from Lenihan. We'd all known him only by his initials; he disdained a *nom de guerre*. Had I ever known his name was Jerry? Yeah, somewhere Jeremiah T., and it had to be; he was the only one in Cholon hospital. I'd visited him the day of the bombing, two days ago but, like the lieutenant, Lenihan was sedated at the time, more out of it even than usual. And flying back-to-back skeds and trapped in yeoman paperwork, I'd not gotten back to that Chinese section of Saigon.

Back at my hooch I peeled myself out of the sodden flight suit, skipped chow while showering, donned civvies, and caught a cart for the front gate. I was slurping down a bowl of U/I meat and noodles by 100-P Alley when I talked an MP into a ride to Cholon Navy Hospital.

Lenihan was propped up in bed, sheets thrown off his immense, hairy body, naked but for skivvie shorts. He was reading a paperback copy of Washington Irving's Sketch Book, some Lit teacher out of his past obviously having an impact. Standing in the door of the two-bed room before entering, I surveyed the wounded sergeant, the most un-military NCO I could name, and smiled at Lenihan's insistence on wearing white military boxer shorts, trashed as decadent by even the most fervent militarist. He was an unlovely sight as he swiveled his head toward me, looking startled. And well he might, I thought, considering his encasement in swathes of white.

"Yo, J.T., Or is it now to be Jerry?"

"What?" The patient leaned forward, cupping hands behind both ears. His forearms were bandaged and, where visible above and below the gauze, the skin was pitted with hundreds of small wounds scabbing over. The bridge of his nose looked to have been chewed by a bad-tempered Chihuahua; it glistened with salve. Both eyes were bloodshot below the cap of white.

I placed a small, brown paper bag on the bedside table and closed the door to the hall. The other bed was empty. Before making conversation I picked up a bottle of rubbing alcohol from the second shelf of the nightstand, poured the contents down the sink drain on the wall, rinsed the bottle, then took a pint of vodka from the brown bag and poured it into the alcohol bottle. I placed it back on the stand and threw the bag in the trash, stuffing the empty Stolichnaya bottle in my pocket. All while Lenihan watched silently, smiling.

It was difficult making conversation: the wounded sergeant could hear only snippets of any string of syllables, his eardrums gone in the blast. I felt awkward, trying to make myself understood without shouting. I had opened the door again and there were Navy medical personnel, military visitors and, likely, VC infiltrators, up and down the hall in a steady stream. I forgot any notion of trying to find out—if J.T. even knew—why he had been brought to the Navy hospital in a distant part of the city, rather than the Air Force facility on Tan Son Nhut, nearer the scene of the bombing. Probably considered to be information I did not need.

"Bob," Lenihan grated loudly, but with an air of conspiracy. I just nodded.

"Bob, you gotta help me. My lockers are full—"

It suddenly occurred to me what the urgent call was about. I leaned forward, finger to my lips, and dramatically, with eyes wide, shushed Lenihan. At the bedridden man's startled look, I mouthed, "Not . . . so . . . loud!" The patient nodded.

"I got weapons in my foot locker," he said more softly, still loud enough to be heard in Vung Tau. "Handguns. Ammo. And four, five long guns in the wall locker." He reached into a small top drawer of the stand and took out a ring of keys. "Here. Keys to everything. Do something with the guns. Sell 'em, trade 'em, give 'em away. Don't care. Keep the money. Just get them out of there before my gear's inventoried."

Army policy dictated that any person, dead or alive, who was not present or accounted for at reveille in his unit was treated as a loss on the morning report, whether a casualty, on R&R or leave, AWOL or a deserter. All personal property of said 'loss' was soon inventoried and stored, or shipped forward, depending upon circumstances. Any Officer delving into Lenihan's lockers would, by mandate, be initiating courts-martial charges on the NCO. Being in a place where war was happening did not, within command policy, convey the right for an individual to bear arms. Certainly not to maintain a weapons cache. Most officers retained their issue sidearms at all times, but enlisted men—even NCOs—were not then accorded that privilege. Later in the war, some rules were changed.

But it wasn't in this case just a weapon, a personal weapon. That might have been overlooked, even if not condoned, the weapon deposited in its proper armory location, the matter forgotten. Lenihan, however, was dealing weapons like some middle-eastern, back-souk entrepreneur. He kept, for his own use, a Swedish-K; he had one more -K for sale; he also had an M-2, .30-caliber carbine, full auto; two Uzis; and an M-14 with scope, ordnance-precision for sniping. In his footlocker, the handguns: two .45 caliber Colt 1911A1s, a .357 Magnum Combat model, a German Luger missing a firing pin, and assorted others I can't specifically remember. But I understood his urgency.

After agreeing to take care of the little problem, as Lenihan referred to it, and about to depart, not anxious to extend the one-way conversation, I pulled the Rolex from my left wrist. Weeks before, my Bulova—bought with my first Marine re-enlistment bonus at the PX, Marine depot, San Diego, eight years before—had conked out on me. It was not waterproof and the festering climate had invested it with moisture, dust, dirt, fungus, sweat, and various microbial invaders until it ground to a halt. I'd mentioned my plight at the table in the mess hall one morning, and recalling my grandfather's bequeathed old railroad pocket watch said I was going to write my wife to pack it up and ship it to me. I squirmed when I thought of the harassment I would endure from everyone: a pocket watch!

But I required a watch for missions and could not afford to buy a new one. Lenihan had offered the use of the Rolex there in the mess hall as casually as he would have loaned me a shoe brush. J.T. came from money, and to him, the watch—one of several he owned, of which he currently favored another with a dazzling display of ganged dials and gears, time zones and bezels, buttons, and buzzers—meant nothing more than a favor for a friend.

Now, when I tried to return the watch, Lenihan's eyes opened wide and, waving his hands, he yelled, "No! You keep it! You still gotta fly. Give it back when you see me again." The not-so-subtle trade-off gift, accompanying disposition of the illicit weaponry, was blatantly obvious.

The mercenary sergeant had dealt in weapons only to offer other GIs an opportunity to safeguard themselves in the quirky but accelerating hostilities; he would not have sold a weapon on the black market downtown. To be sure, however, though never to be acknowledged, he was operating a black market operation of a sort, all his own.

In the next several days, with the delighted conspiracy of a Green Beret from the 400th SOD, every weapon in J.T.'s lockers was moved from Davis Station compound, disappearing into the shadowy world of indigenous support—all except the .357. I found a niche for that in my own locker, looking over my shoulder as I stowed it away. The

Green Beanie paid me a pile of local currency, which I didn't bother counting; but later, through various enterprising exchanges, and at a loss converted it back into American currency. I used the laundered funds to buy money orders, which I sent on to J.T.'s Petaluma home address . It was some weeks before I dropped the last envelope through the mail slot, sighing with relief at the end of the threat to Lenihan ... and to myself.

After SSG John Moore had returned from the PI and the subject of the weapons trade operation was known to him, he said, apropos of nothing: "I suppose it'll be rum-running next."

When Your Number's Up
By George Murphy

In 1966, I spent the summer with the 1st Brigade, 101st Airborne Division. This was my second tour in Vietnam. On my first tour I was at the 3rd RRU. On this second tour I was assigned to the 313th RR Battalion out of Nha Trang. I was sent TDY to Detachment 3 of the 3RRU, later called the 406th RRD. It supported the 1st Brigade of the 101st Airborne Division.

Initially I went to Phan Rang for training. Mostly it was training on firing the M-16. After the training I went to Strike Forward, which was located at Dak To. The Brigade was involved in Operation Hawthorne and there were numerous NVA KIA's. After the operation finished, Det 3 moved with the Brigade to Tuy Hoa.

In addition to the standard Morse intercept, the detachment implemented a program of low level voice intercept. ASA personnel would go into the field with infantry troops on combat operations looking for VC/NVA voice communications. On Operation John Paul Jones, I went into the field with the 2nd Battalion 502nd Infantry, nicknamed Gunfighter. There were two ARVN's with me. They were the ones who listened to the radio. There was also a second ASA guy and for the life of me I cannot remember who it was. While I did this several times, we never really heard any enemy voice communications.

For John Paul Jones we were trucked out of the Tuy Hoa Fire Base one morning in August. After a short ride, we offloaded. We would walk in the rest of the way. After the deuce-and-a-half ride, we were informed that the road was mined. We were instructed that we should only walk on the paved roadway. After we passed, a bomb disposal team would clear the road. OK, why clear the road after we passed? I have no answer.

On this operation the 502nd was relieving a battalion of ROK Marines. As we walked in, they walked out. They were all smiles and waves. The two ASA guys and the two ARVN's were at the end of the column of several hundred soldiers. I can report that no one stepped on a mine.

After what I think of was several hours of marching, we arrived at a deserted village. It looked like part of it was burned out. On the edge of the village there were expended napalm canisters. You could see damage from earlier action. There was one masonry building and several huts with thatched roofs. There were also several animal shelters. The battalion HQ and the aid station took over the masonry building. The two ASA guys and their ARVN's were assigned to one of the animal shelters. It was an open structure with a thatched roof. We were instructed to dig foxholes, where we would spend the night. After we dug through six inches of animal dung, we found dirt.

Most of the 101st soldiers ignored us. I heard us referred to as a tag along. I do not think they knew why we were there. At the time I do not think anyone with the 101st had a high enough security clearance to know what we were doing. I would even include the battalion CO.

There was only light contact with the VC and there were no wounded in the aid station. For whatever reason, a medic seemed to befriend us. While the others ignored us, he came over and talked to us. It might have been he was bored or just curious as to what we were doing.

After we had dug in for the night, the medic came over and told us there were C Rations at the HQ and that there was a small stream on the other side of the tree line. If you needed to fill your canteen, that was the place to do it.

I thought I would fill my two canteens. I walked across an open area behind the HQ and through the trees. I came out at a small stream that was less than a foot deep. The stream meandered. On my side there was a small sandbar. On the other side the brush came down to the water. I took off my web belt and combat pack along with my steel pot. I rested my M-16 on my steel pot. In my pack were some low level classified documents. I do not think anything was higher than confidential. Attached to my back pack was a white phosphorus grenade (Willy Peter). I was told if I got into a bad situation I should pull the pin on the Willy Peter and run like hell. Since Vietnam, I learned that WP burns at 5000 degrees. Pulling the pin would have been interesting since in addition to the classified material, some dry socks, and leftover C Rations, I had at least 300 rounds of M-16 ammunition. I did not have to pull the pin.

I took my two canteens and filled them from the stream and then I dropped iodine tablets into the canteens. I gave them a little shake. I returned the canteens to their carriers on the web belt. It was a pleasant sunny afternoon. I decided to light a cigarette. I sat down on the sandbar to enjoy my smoke.

After a few minutes to my surprise, a VC came out of the brush on the other side of the stream. He was about 30 yards up stream. He was dressed in black pajamas and a straw conical hat. He was carrying an AK-47 with a bandoleer across his chest. He looked at me and I looked at him. My M-16 was at least an arm's length away. I had no cover. He did not raise his weapon and fire. After several seconds, which seemed much longer, he turned and headed back into the bush.

I grabbed my stuff and ran back to the battalion HQ. There was a sergeant sitting on the front step of the HQ smoking a cigarette. I explained to him what had happened. He did not seem concerned. He told me that based upon information from the ROK Marines, they knew there were VC stragglers in the area and that the security platoon was setting up our night perimeter. He said he would pass it along.

I walked back to our animal shelter. The ARVN's were setting up the radio that they would monitor. I sat down on the dung floor and lit another smoke. I did two packs a day back then. I do not think that the Army gives out smokes anymore. After a few minutes, there was a pop, pop, pop. A weapon was fired. We all turned and looked at the HQ. There were a dozen soldiers sitting on the side of the building. They never moved. The HQ was quiet. So we just sat there.

A little later the medic, who befriended us, came by. He said the security platoon had fired at a VC who got caught inside our perimeter.

I do not know if the VC I saw earlier that day was the one fired at. Who knows? All I know is that on that sunny afternoon in August, 1966, just short of my 23rd birthday in a stream bed, next to a no name, abandoned, and burned out village, in Phu Yen province, not far from the town of Tuy Hoa, my number was not up.

Four plus decades later, I sometimes think about the events of that day. I think how things might have gone. I think that the VC saw the ROK Marines leave the area and decided to take a look. He was not expecting to find the 101st. When the VC saw me, he wanted out of the area. He did not want a fight and I do not know if I could have given him one.

A Tactical SIGINT Success Story
By LTCOL Steve Hart (Ret'd)
with
BRIG Ernie Chamberlain (Ret'd)

Prologue - *A little known battle of the Vietnam conflict:*
547 Signal Troop (hereinafter, the Troop) was a sub-unit on the Order of Battle of the 1st Australian Task Force for the duration of the Vietnam War. Its role was to provide timely Signals Intelligence (SIGINT) to the Task Force, initially operating only as a 'post office' for the US Army SIGINT organization, but quickly becoming a significant operational SIGINT resource in its own right. [1]

In an era before the Australian Army embraced Electronic Warfare (EW), which did not occur until the late 1970s (and strongly motivated by the Troop's successes in South Vietnam), the Troop was required to operate under stringent strategic security regulations applying to the existence and distribution of SIGINT. This necessitated restricting the access to its end-product to a very small number of appropriately-cleared recipients, many of whom, in the early days of the deployment had very little exposure to, or confidence in, the intelligence product being developed. This, together with the 'fog of war', was certainly the case in the lead-up to the Battle of Long Tan. As a direct result of that battle, the Troop strength was doubled from 15 to 30 and an R & D task (Project HIGH DIVINE) was initiated with the Weapons Research Establishment (WRE) to equip the Troop with its own high frequency airborne radio direction finding (ARDF) equipment, to be installed in the fixed-wing aircraft of 161 Reconnaissance Squadron. Officially, the operations of the Troop continued to be 'highly classified' as secret plus until the release of the three volumes of the official history of the Australian Army in the Vietnam War which contain numerous citations referring to the contribution of the Troop to Task Force operational planning.

In April, 2012, the *Defence Honours* and Awards Appeals Tribunal initiated an "Inquiry into the Recognition for Service with 547 Signal Troop in Vietnam from 1966 to 1971" to examine relevant evidence and consider appropriate recognition for the Troop's contribution. A number of supporting submissions were made including a detailed Troop submission. This article expands on just one engagement reported in that submission – one that has yet to be recognized in any official record of the War.

Much of the prose in this article has been based on the recall of Troop members who were directly involved in the incident. The authors are particularly indebted to Adrian Bishop and Jeff Payne for their contribution. The italicized prose is based on detailed historical research of North Vietnamese and Viet Cong records and unit histories undertaken by Ernie Chamberlain and focuses on the enemy view.

The Strategic Situation in May-June 1969

The communist headquarters in South Vietnam ordered a month-long campaign of 'High point" attacks for the period 5 May-20 June 1969. Following the late May announcement of the planned early-June meetings of Presidents Nixon and Thieu at Midway and the communists' planned announcement of the formation of their Provisional Revolutionary Government, further significant attacks were planned. With the occupation of Binh Ba village by elements of the 33rd NVA Regiment on 5 June 1969, the Australian Task Force's 5 RAR became involved in heavy fighting over several days in both Binh Ba (Operation Hammer) and Hoa Long (Operation Tong) villages.

]274 VC Regiment – Plans Uncovered

In mid-June, 1969, during normal radio frequency search operations, two experienced Operator Signals of the Troop, Corporal Roy Dean and Signalman Barry Nesbit, recognized - purely by its aural characteristics, the radio transmitter of the Headquarters of the VC's 274th Main Force Regiment which had been observing radio silence for several days. While Dean and Nesbitt copied the radio messages, the target frequency was also being monitored by Troop member Signalman Jeff Payne, who was flying a scheduled ARDF mission in a Cessna 180 from 161 Reconnaissance Squadron, piloted by Lieutenant Tony Sedgers. Sig Payne undertook the tracking procedure and immediately passed this data (via a secure voice link) to Sergeant Bob Hartley who performed the appropriate calculations and plotting to determine the location of the target transmitter. This indicated that the radio station serving the Headquarters of the 274th Regiment was located near a Thai Army forward base in Long Thanh District of Bien Hoa Province, 38 kilometers north-west of the Australian Tactical Area of Operational Responsibility (TAOR) but within the Australian Tactical Area of Intelligence Interest (TAOI).

One of the Troop's cryptanalysts-linguists, Corporal Adrian Bishop, was able to decipher and translate the intercepted VC radio messages. The messages said, *inter alia*, "When the combat is over you are to evacuate the wounded to the hospital in the May Tao Mountains as agreed at our planning meeting." This single piece of intelligence indicated that a significant attack was imminent as the Regiment's casualties were usually evacuated to the nearby Hat Dich Secret Zone.

The information was quickly passed up the chain to Bien Hoa, thence on to the Special Security Office serving the Headquarters of 2 Corps, Field Force Vietnam (II FFV). After several urgent exchanges between intelligence staffs, it was agreed by all sides that the most likely target was the battalion base of the Royal Thai Army Volunteer Force (RTAVF) 2nd Infantry Battalion/1st Brigade/Black Panther Division located at Loc An - about 3.5 kilometers south-east of Long Thanh District town and twenty-eight kilometers east of Saigon. This Thai position (2) was defended by two companies – numbering 245 personnel, and included a US SNCO as a Liaison Officer. Arrangements were quickly made to harden the defenses of the Thai position and prepare for the attack.

The Attack

At about 1.00 am, on 16 June, the VC Commander – Nguyen Nam Hung - launched his attack on the Thai perimeter with two battalions plus sappers and one battalion and RHQ in reserve. A barrage of VC mortars and Rocket Propelled Grenades was followed by three waves of ground assaults and greatly outnumbered the Thai defenders. However, the attackers were taken completely by surprise and were repulsed with a bewildering amount of firepower including small arms and mortar fire, artillery support, claymore anti-personnel mines, helicopter gunships as well as other US close air support – including napalm. The battle raged for several hours but the defensive fire was so intense and casualties so high, the VC Commander was forced to call off the attack and retreat.

By coincidence, the Troop's Senior Traffic Analyst, Staff Sergeant Darryl Houghton, was visiting a fraternal American intercept organization unit at

Bien Hoa and was due to return to Nui Dat later on the morning of the attack. The Americans were so impressed by the intelligence tip-off and the successful defense of the Thai position, they diverted Houghton's helicopter to the site of the battle. Houghton arrived just as the body count was completed; 212 enemy bodies were found. The fire from the helicopter gunships and the base defenders was so intense that the VC was not able to drag away the bodies of their KIA as they routinely did during other battles. [3] Thai casualties in this battle were six killed and thirty-four wounded.

In subsequent notifications [4], the 1ATF Intelligence staff reported that the action 'resulted in 212 enemy KIA and one enemy PW ... Numerous individual and crew-served weapons, ammunition and other materials were captured. Captured documents identified the elements of all three battalions of the 274 Regiment, K21 Sapper Reconnaissance Company and rear elements of 274. This assault ... has probably rendered the Regiment marginally effective as a regimental-size combat force. It is assessed that remnants of the Regiment have withdrawn to their Hat Dich base area and will probably remain there for at least a month, until replacements are received'.

An important factor to note here is that the number of VC casualties incurred in this battle in June, 1969, was almost as high as those suffered by 275 Regiment at the Battle of Long Tan with the Australians almost three years earlier.

Subsequently, and based on the Troop's information, the Task Force Commander (Brigadier Sandy Pearson) directed ambush action on the expected 274 Regiment withdrawal routes eastward to the May Tao base area. Successes were reported a few days later. [5] "Early am on 20 June 1969, 6RAR elements ambushed a 50-strong VC group moving from west to east, killing twenty-two at YS498896 [east of Route 2 and just inside Phuoc Tuy Province] – captured documents tentatively confirmed the 2nd Company of 274 Regiment's 1st Battalion." A VC POW confirmed that the group had been engaged in evacuating wounded from 274 Regiment following their unsuccessful attack on the Thais. While the ambushing elements of 6RAR had been positioned there on the basis of the Troop's earlier intercepts, this is not acknowledged in any history. Hence the excellent teamwork resulting from this one intercepted encoded message, and the accurate ARDF fix, resulted in 234 enemy KIA in just four days.

Yet the background behind the attack on the Thai position in June, 1969, does not appear anywhere in the annals of Australian (nor apparently in Thai) military history. [6] *Perhaps understandably, the failed attack on the Thai position is not mentioned in any published Vietnamese Histories of the War, including in the history of 274 Regiment's higher headquarters – the 5th VC Division and Military Region 7, nor in the recent memoir of the 274 Regiment Commander, Nguyen Nam Hung – now a retired Major General. Further research of Vietnamese documents reveals that the C.12-65 Binh Gia Assault Youth Unit from Phuoc Tuy Province was involved in portering the 274 Regiment casualties eastward to the May Taos. And there was a brief Hanoi account in an article titled: 'Dazzling Military Feats during June' and dated 1 July 1969 – 'On 15 June, the PLAF of Bien Hoa Province destroyed a battalion-size unit of Thai troops near Long Thanh'.*

It is almost certain that the recipients of the 'tip-off' in the Thai perimeter that night were never told that the source was from an Australian intercept. The story is only known to a handful of people outside of the Troop, mostly Americans in the SIGINT community.

Lessons Learned

Tactical SIGINT is timely and potentially 'actionable', but extremely perishable as events move quickly. Quite often, it is not possible to verify this intelligence by other means. The above incident is just one example where the Troop expeditiously passed information to relevant combat forces needing the information. Their timely and effective reaction to that SIGINT turned a potentially disastrous situation into a highly successful outcome.

Epilog

At the time of writing this article, the deliberations of the *Defence Honours* and Awards Appeals Tribunal are still in progress. The achievements of this small RASigs unit have effectively been suppressed for over forty years - primarily in the perceived interests of national security. However, it is at least gratifying to the Troop's members that just some aspects of their contribution to Task Force operations have now been acknowledged in the Official Histories.

About the Authors

At the time of this incident, Steve Hart was the Officer Commanding 547 Signal Troop and Ernie Chamberlain was the Intelligence Liaison Officer in Baria.

Endnotes
1. Issue 57 (Summer 2012) of Wartime, included an article entitled 'Listening to the enemy. Signals Intelligence played a key role in Malaya, Borneo and Vietnam', Dr. John Blaxland briefly reported on the contribution made by 547 Signal Troop in the Vietnam conflict.
2. Commanded by LTCOL Mahitorn Booyanurg who had commanded the battalion in Vietnam for almost 12 months.
3. The enemy also left over 824 grenades – i.e.,

  sapper assault weapons but recovered all but
  16 AK-47s, 18 RPGs and two SMGs.
4. 1ATF INTSUM No. 167/69 of 16 June 1968
  with a more detailed report in 1ATF Vietnam
  Digest No. 24-69, 14-21 June 1969.
5. In 1 ATF INTSUM No. 171/89, 20 June 1969.
6. A brief description of the attack, based on
  research by military historian Dr. Bob Hall of
  the UNSW at the Australian Defense Force
  Academy, and Air Vice Marshall Vorachat
  Tharechat and Group Captain Sakinit
  Promthep, was published in the Newsletter of
  'Australian-Thailand Association (Canberra)
  Incorporated', October-December 2010,
  Volume 36 Issue 5.

Chapter 5
ASA in Other Places

On Another Front
By Robert Flanagan

December, 1960. Christmas approaching, and all
through Ops the sound of carols drowns out
ambient dits and dahs and white noise. The
nostalgic warmth of the season overrides the fact
that here we are, stuck on an 8,000-foot-high
plateau on the northeast edge of Africa, bordering
the Red Sea: the 4th USASA Field Station, Asmara,
Eritrea. Not just on Kagnew Station, but in
downtown Asmara, the nativity spirit prevails.
Ethiopia, which rules Eritrea by UN fiat at this time,
is a Coptic Christian country; and, too, there is
ample Italian Catholic influence from their long
history in Ethiopia. Other religions abound,
including many Muslims, animists, and off-shoot
sects of indescribable complexity. But Christmas is
coming!

Working 'swings' on the evening of the 13th,
no one was thinking of anything more than the end
of the trick, maybe a Melotti or two down at the Blue
Nile Bar, or even a short session with one of the
'Annie' bar girls: Big Annie, Dirty Annie, Lost Annie,
Cockeyed Annie, Little Annie, and Lottie.

Just before midnight, a comm-center corporal
sidled down the aisle in Bay one of the Manual
Morse section, dispensing his own brand of
excitement.

"Hey," he looked around cautiously, and we
all knew we were about to be privy to some
indiscreet information, "just got a FLASH from the
embassy in Addis. Some shit's hit the fan. Don't
know what."

"What do I care, *corporal*; I got all the
classified chatter I can handle right here."

"Yeah, go back to your vault, Simmons. Gimme a break." Catcalls and dismissals arose across the humming bay. Disdain flowed like Tuborg, and the loose-lipped corporal scampered away to seek more receptive ears. Since we didn't copy host country, we had no clue what was happening in-house. Didn't much care: Santa Claus is coming to town."

By the 14th, though, the comm-center circuits were overloaded, spilling FLASH and CRITIC messages like ticker tape of another era, though the workers-behind-the-grate had closed ranks and there were no further indiscretions. But a buzz was wafting about the post, and by the time we went back on swings that day, target circuits in surrounding countries were cranking out top-priority traffic by the ton. The consensus was: a coup! A coup d'état, in the making.

The trick chief, an SFC who played the 'knowledge is power' card, fed us little snippets reminding us that Emperor Haile Selassie was out of the country on a state visit to Brazil. The minuscule autocrat spent a lot of time flitting about the world, pursuing some elusive agenda. Now would be the time, all right, for malcontents to make such a move.

And the rumor mill was right on: A coup, at least in the making, had been crafted in the capital of Addis Ababa, far to the south of us. Eventually we learned that a provincial governor, who was anti-Haile, and his brother, a general in command of the Imperial Bodyguard, had led a small group to snatch power, initially capturing and holding hostage a number of important government ministers. The emperor's son, Crown Prince Asfa Wossen, made a radio broadcast, announcing that the old government (Haile and crew) had been deposed, and he was declaring a new, progressive regime.

It was later determined that the Prince had a gun to his head when he made his pronouncements. The rebels captured the usual suspects—the main radio station, the airport, military headquarters in the capital, and the beer supply—but they failed to lock up some critical alliances that might have enhanced success: the Imperial Tank Regiment was still on the loose, and fiercely loyal to Haile; and the Imperial Air Force, who flew their F-86 Saber jets over the city, breaking the sound barrier to impress stone-age residents.

Another imperial son sent a message on the situation to his father in Brazil, using his HAM connections. With the airport lost, Haile could not even fly back into his capital; so he flew into Asmara, and drove in an armored column from our mountaintop to Addis Ababa, escorted by his 2nd Infantry Division, veterans of the Korean War and fiercely loyal troops who were in Eritrea acting as national police during the mandate period. The population of Ethiopia was poorly educated, and suffered entirely from lack of information. They were told Haile was deposed, which automatically meant 'dead,' and they went with the flow. But when his column drove into Addis, the emperor standing erect in the lead vehicle, every person in the realm dropped on his face and the coup began to unravel. This latter step happened on the 17th of December.

But in that interim period of frenzy, when no one—apparently in our government or theirs—knew what was happening, strange and exciting events were manifest. At Kagnew Station, armed MP Jeeps accompanied the buses which transported trick workers from the billets out to the Ops site, and back. NCOs sported side arms; a number of other troops were carrying 'discretionary weaponry', i.e., illegal and unaccounted for.

The Commander, an Officer who had the rank but garnered no respect from anyone conscious of his presence, left the O-Club after having a few, directed his personal bodyguard MPs to get him to the motor pool where an old tactical armored vehicle was dusted off, cranked up, and with the CO aboard, clanked and ground off down the macadam streets, playing hell with the tarmac, and out the front gate to the astonishment of both American and Ethi guards, disappearing down the hill toward town. Firing was heard down the hill later—automatic weapons fire—and sphincters tightened. Next day word trickled down that the DUI Colonel had opened fire on the walls of several homes in the town with the .50 caliber mounted on the APC. Those rumors were quickly shut down as if by a vow of *omerta*; I could never confirm the firing.

But on the next to last night, before anything was resolved, we on swings looked up to see that same comm-center corporal charge into the bay, eyes bugged out, yelling, "The Russkies are coming. They're *en route* now; gonna drop paratroops to take Kagnew Station." Startling information, when the most exciting thing we'd had up until now was a distant bit of flak taking place in the Congo. I can't say how many people bought into this scenario, but naturally, things got a bit antsy. Positions were temporarily abandoned, as nervous ops tried to find someone with the straight scoop. There was no such.

By then it was known that there was an element of communists or socialists involved in the coup. And this airborne assault hysteria occurred before Haile returned, turned off the coup switch, and shot and hanged a bunch of rebels. As in most field operations, we never got much in the way of feedback on the entire production. There were rumors of a few deaths in Asmara, many more down country. The Organization of African States later put the death toll at above 2,000.

To the corporal's startling announcement, there were responses like, "Hell, let 'em come. I'll make sure the coffee's hot for 'em."

"I'll leave my R-390s to them in my will," and assorted other weak attempts at humor.

But by and large, there were anxious soldiers in the Ops building. No one ever tried to explain what a remote American intelligence collection site, some 900 miles away, would have to do with the overthrow of the emperor. Of course, by the time the 17th rolled around and Haile was back in charge, the fears quieted. No Ivans dropped in on us. Some doubted they could put paratroops on that plateau anyhow, seeing as how we often couldn't find our way back to the barracks from the Oasis Club, right there on the same base.

It took another 14 years for 'them' to get Haile, bring down his long-reigning house, and set the nation into 30-plus years of civil war and revolution. Even after ASA abandoned its site on the plateau, NSA hung around for years with PROJECT STONEHOUSE, until rebel forces gave reason for that to die a gracious death, too.

But back in 1960, that Christmas came and went. A fond memory I have of Christmas day is that NCO and officers' wives came and set up shop in the Day Room of the barracks, and dispensed coffee, lemonade, fruitcake, cookies, fruit and other goodies. Made a nice impression on a SP-4 far from wife and child, and promised better for 1961.

But as the New Year rolled over, another flap developed that was linked back to errant comm-center types again. It developed ... then quickly faded. It was another corporal who triggered this one, too. Still on swings, we were then into a ten-month stretch of working 'straight tricks, where we didn't rotate, but worked the same trick consistently.

It was a Sunday night. I was by now an assistant trick chief; an A-Slash in the vernacular of the moment. A comm-center weenie rushed into the Morse bay: "Hey, A-Slash, pass the word. Headquarters going to be announcing tomorrow that the Hall needs volunteers to go to Indochina."

"Say what?"

"Godstrewth! Indochina! Don't know much about where that's at, but it's something different."

He was ecstatic to have found an alternative.

Well, the A-Slash wasn't the only one who heard the breathless news. By the time Mids Trick came on, the entire Ops building was ablaze with discussion. I knew a couple of things about Indochina: that's where Terry & The Pirates hung out their shingle, and just about six years before, there was a rather injudicious conflagration at a place called Dien Bien Phu in Indochina.

I was not in the queue at the door to the S-1 the following Monday morning ... but some forty-or-so other 058s and 982s were lined up, clamoring for the adventure of a lifetime. The line formed long before the shop was even manned; some of the would be applicants had staggered directly to the Head Shed from the Oasis Club, when it closed at 0400.

The Personnel Sergeant, a SFC without a sense of humor, was the first to arrive. You can imagine his state-of-mind when he was faced with a swarm of avidly pleading applicants, and he hadn't even seen the morning Read File, didn't know what they were going on about.

Of course, the classified message, I recall it was SECRET, maybe with some caveat, was as good as in the morning newspaper. America's hatchling interest in Diem's little operation along the South China Sea was close-hold no longer. After a bunch of 'arrests'—these consisted of many of the lineup being told to go to their billets, stay there, and keep their mouths shut— they were all directed to return to Ops with regularly scheduled work tricks. Within two days, another FLASH from Arlington Hall was posted conspicuously on the Ops bulletin board, informing all that the USASA had no further need for volunteer 058s and 982s for duty in a Southeast Asian country. That project had been canceled. Any information relating to an unfortunate misunderstanding of recent notification was to be kept in absolute secrecy.

It would take another three years for me to work my way to the front of that queue and check out for myself all the haunts and hauntings of Terry and his erstwhile pirates.

Mrs. Halopowski's Ruckus
& Other Fine Moments
By Don Collins

It's interesting what I remember from my time in the army. Certain moments stand out; there was a handful of scary moments from Nam, and the times on leave back in the States when I felt like an outsider and even remember being called a baby killer for being in the service, and I remember some poignant memories from Teufelsberg, Germany, including a few run-ins with the East Germans and the odd incident involving the unforgettable Mrs. Halopowski.

When I returned from Vietnam in May, 1967, after spending 26 months in SE Asia, we landed at McGuire AFB at 3:00 in the morning. It was about 45 degrees outside, and when they opened the door there was a wave of screams that started in the front of the plane and rolled all the way to the back. NO ONE had a coat or a jacket.

By the time I got to the Philadelphia airport it was up to 70, and at least bearable. Walking through the Philadelphia airport to catch my plane to Ohio I passed a cute young girl who welcomed me home by calling me a baby killer. Having been overseas for two years I couldn't figure out why she would say that, but I learned the answer over the next few days.

My home at the time was Sandusky, Ohio on the banks of Lake Erie, where in late May the temperature is usually in the 70's during the day, and low 60's at night. The locals all thought that the weather was wonderful, while I was freezing to death, and wearing sweaters, and getting weird looks. After about a week of this they had a heat wave with the temperature getting up into the low 90's. Not a lot of people have air conditioning in northern Ohio; you really only need it about one week a year, so the locals were all dying of heat while I was happily enjoying the pleasant weather. I didn't score too many points that week.

After a 45 day leave, I went on to the 54th ASA Special Operations Command in Berlin, Germany, arriving in early August of 1967. There I spent ten months up on Teufelsberg Mountain doing routine maintenance work. The highlight of this period was when one of the guys figured out how to watch TV on an oscilloscope. The only time we could get away with this was on the night shift, and there weren't any West German TV stations operating then. The only thing we could pick up was the East German stations, and all they were showing were old WWII movies. So we would sit there at night watching lots of little green Russians, waving lots of flags, charging and killing lots of little green Germans, who didn't have any flags; probably why they lost.

Several weeks after I got there was the Oktoberfest. The guys were having a lot of fun with it and as part of the festivities a lot of them had these little plastic cap pistols that were very realistic looking copies of a .38 snub nose. I was aware that they were horsing around with them but never really paid any attention to it.

One day I was walking down the hall toward my room and there were several other people in the hall. As I got near my room my roommate whipped out his .38 and pointed at me. I reacted, immediately diving for cover, and I remember flexing my right hand where an M-16 normally would have been. I looked around and everyone was just standing there in shock. Remembering where I was, got up, and continued to my room with everyone's mouth still open.

Never, ever, saw a cap pistol after that. Also, I had a new roommate a few days later.

One of the operators on the hill was a timid little guy with coke-bottle glasses named Halopowski. His wife was so huge that she actually had a kid, and no one knew that she was pregnant. She also had a horrible temper. It was so bad that she would actually terrorize everyone around her everywhere she went, with her timid little husband following along apologizing to everyone.

Well, her husband had been pulling KP about three times a month, which was more than his fair share. Turned out that he had requested the KP because it gave him more time away from her, but she didn't know that. She was mad about all the KP he was pulling and went to the First Sergeant's office to raise hell about it. The First Sergeant managed to escape and ran and hid in the NCO club.

She then went looking for the Commanding Officer. The CO's office was in a secured area on the second floor of the Headquarters Company which was guarded by a very small MP. If you wanted in you had to knock on the door, which he would open, and ask what you wanted. So she went up there and banged on the door. When the little MP opened the door she slammed the door wide open with her right arm, and with her left sent the MP sailing down the hall. There was a Com Center there, and when the guys in the center saw the MP flying by their door they thought that they were under attack. They put the appropriate tape in the machine, grabbed their M-14's and went charging down the hall. When they got to the CO's office they found Halopowski's wife pounding on the desk and screaming at the CO, who was trying to get out the window, and not having much luck because it had bars on it.

Of course after sending that tape, Frankfurt wanted an explanation of what was going on. Shortly after this she was sent back to the States, and Mr. Halopowski's morale improved considerably.

Thinking that they were under attack wasn't too far off the wall. They were mildly concerned about Russians trying to get into Andrews Barracks. There were teams from all countries that went into each other's zones. I remember from my CQ duties that there was a written contingency plan on what to do if one of these teams entered the compound. Having just come from Vietnam I thought the whole thing was rather quaint.

Right after this they re-engineered the door to the second floor. The room next to the stairs became the guardroom. They cut a hole in the wall for a window so that you could talk to the guard. To let you in the MP had to push a button on the wall that unlocked the door. There was a big heavy steel plate between the window and the button, so that the button couldn't possibly be reached from the window.

Never saw the small MP again after that. My room was two floors above the secured floor and I went up and down past the door on a regular basis. Sometimes the door would be open and the MP and I would talk for a few minutes. He was a nice guy, and I hope nothing bad happened to him.

On a side note, my Maintenance Officer, Mack Criddell, *who was black*, ran into Mrs. Halopowski in the mailroom one afternoon. She proceeded to lecture him for several minutes about how hard it was being a minority, and that he could never possibly understand what it was like to be discriminated against! Mack retired to Dallas, and over the years I would run into him every once in a while. After 45 years he was still pissed off about that!

By some weird twist of fate, the unit had a very high number of people that had college degrees in teaching. Whenever the treads, that weren't as well educated, would put an order on the board, they would proceed to grade it for grammar, punctuation, spelling, etc. and give it a letter grade. This drove the treads absolutely nuts! They finally put out an order that there would be no more grading. Of course that order received an "F".

We had a forty gallon coffee maker up on the hill, which produced what everyone considered to be the world's worst coffee. Periodically we would have drills in which we would have to simulate the destruction of the site. This was accomplished by hanging "Destroyed" tags on the appropriate equipment. Whenever the alarm went off, people would kill themselves trying to be the first one to hang a Destroyed tag on the coffeemaker. Only when the coffeemaker had been "Destroyed" would they start in on the rest of the site.

One day a new guy went and got his first cup of coffee. He came back into the shop, and took a sip. His faced immediately puckered up, and he exclaimed, "This coffee is horrible! It tastes like there's a dead rat in it!" One of the other guys, hearing the comment, picked up his cup, took a sip, swished it around, thought about it for a moment, and replied "No…I think they took the rat out."

When the Berlin Brigade had an alert, they would actually deploy, and try to hide in the Gruenwald, which is about like trying to hide a brigade in New York's Central Park. My roommate was assigned to the team that went out with them. There was never anything for him to do during these exercises, and he tended to hang out with the Physiological Operations team, which also didn't have anything to do. One time they ended up in the woods next to a section of the Wall dividing Berlin. This particular section was all barbed-wire and with dogs patrolling in between the fences. The Phys Ops people had a Jeep with them that was designed to talk the enemy into surrendering from about ten miles away. The whole front end of the Jeep was a huge speaker system, which was powered by a gazillion watt amplifier. Having a lot of time on their hands, they decided to hook Steve's signal generator up to the speaker system. They set it at a high frequency that only the dogs could hear, and turned it loose. The dogs went wild! The East German guards came out to see what the matter was, and were bitten by a few of the dogs for their trouble. The East Germans finally retreated, while our guys were left laughing in the woods.

Berlin was the end of the line for me. I returned at the end of May, 1968, and was discharged at Fort Hamilton, New York.

Monica
By Don Collins

Toward the end of my tour in Vietnam I had to fill out a dream sheet telling the ASA where I would like to go next. I put down Homestead, Florida, Two Rock Ranch, California, and Vint Hill Farms Station, Virginia. Naturally, I got Berlin, Germany, with only ten months left to serve.

Our barracks were at Andrews Barracks, which had a history going back to at least the beginning of the century. It had been the training center for Kaiser Wilhelm's Kadeten Corps, and later, Hitler's SS Officer's barracks. It also had a swimming pool used for the 1936 Olympics.

Across the street from the gate was a bar that looked like it had been there the whole time. It still had its well-worn blackout curtains in place and ready to be used at any time. The old guy that ran it looked like he had probably sold beer to Kaiser Wilhelm's Kadets.

One of the regulars was a lady named Monica. Sometimes she was a customer, and sometimes she was the bouncer; it depended on the current needs. Monica was almost six feet tall, weighed at least 300 pounds, and had arms that would have impressed Arnold Schwarzenegger. Because of her size she had trouble meeting guys, and was frequently depressed about her sex life, or lack thereof.

We had had a guy transferred into our unit a few weeks earlier. One night he was at the bar and got into it with Monica over something when she was in bouncer mode. The disagreement was settled when Monica picked him up and threw him out an open window. The bar was about six feet above sidewalk level; you went upstairs to get to it, but fortunately he landed in a hedge, suffering only minor scrapes.

A week or so later we were bar-hopping and decided to stop in there for a beer. Monica was there as a customer this night, and trying to pick up a guy for a roll in the hay. She came over to our table, sat down, and proceeded to try and talk my co-worker into going to a hotel so that she could make amends for throwing him through the window. All he needed for a night of wild abandon with her was ten dollars for the hotel room. He said that he would like to but he was broke and didn't have the ten dollars. She was sure that he could find the money. This went on, back and forth, for about a half hour with me sitting there not saying a word and drinking my beer.

Eventually nature called, and I went to get rid of some of the beer. When I came back and sat down, Monica leaned across the table and growled right in my face that my friend said that I had ten bucks and she wanted it or else! Not wanting to end up in the hospital, I pulled ten bucks out of my wallet and put it on the table. She stood up, grabbed the money with one hand, grabbed my about to be ex-friend with the other, and tucked him under her arm. As she walked out the door with him under her arm his last words were a screamed "I'LL GET YOU, COLLINS!" It was about this time that I realized that she had bluffed me.

The next day when he got back he staggered into my room, looking considerably worse for wear, and told me that he was going to give me my ten dollars back in the bar right in front of Monica. I took that very seriously, and stopped going into that bar. Fortunately, for me, he was transferred to Vietnam a few weeks later. When I did go back to the bar and ran into Monica she asked me where the 'Jackrabbit' was. I got the impression that referred to his performance in bed. She wasn't too happy about it.

There Was No Radio Research In Viet Nam
By Ed Johnson

I was discharged in January, 1970, after spending two tours with the 265th Radio Research Company with the101[st] Airborne Division. I graduated from The University of Dayton in 1973 with a B.S. in Business Administration.   After interviewing with a number of companies I accepted employment with a local accounting firm in Chicago.

Shortly after starting with this company it became obvious that they had severe management, organizational, and leadership problems. After three or four months I started saying to myself that I wished I was back in the Army.

In the early to mid-70's the economy was in recession and jobs were hard to come by. I began actively searching for a new job and after several unsuccessful months put my career change on hold. The conditions became unbearable due to internal power plays, squabbling, and backstabbing.

One day, after meeting with a client, I walked past the local Army Recruiting Station and saw the recruiter sitting at his desk and he was obviously bored. I thought I would brighten up his day and sat down in front of his desk. I gave him my background and he said with a college degree I could enlist for the OCS option.   He was very forthright and gave me an information packet on what was available.

I went back to the office and my section watched several of our coworkers get publicly skewered for acts we knew they didn't do. I reviewed the recruiter's packet and gave careful consideration for returning to active duty during the next several weeks. One day after a week of bullshit, I called the recruiter and set up an appointment for testing and a physical exam.

After testing, physical exam, and signed OCS option, I executed the Johnny Paycheck option and told my boss to *Take This Job and Shove It*. I was off to Fort Polk for basic and AIT. From there I was sent to Fort Bragg to the 82nd Airborne for jump refresher and to wait for a slot in the next Infantry OCS class. At that time there was only one OCS Company with two classes a year. After 3 or 4 months I received orders to Fort Benning and Infantry OCS.

After graduating from OCS and Ranger School in early 1976, I was sent to Fort Sill for the Field Artillery Officer Basic Course. Fast forward to 1982-1983; I was commanding a self-propelled 155mm artillery battery with the 1st Armored Division in Nuremburg, West Germany.

Late one Friday afternoon we were gathered at the Nuremburg Officer's Club for happy hour. I was standing with a group of mostly captains and lieutenants in the Division when one of the brigade commanders stepped up to the bar to order a drink. He obviously saw my 101st patch on my right shoulder.

"What unit were you in with 101st?" he asked.

"The 265th , sir."

"The 265th what?"

"The 265th Radio Research Company sir."

"I spent two tours with 101st and never heard of this 265th Radio Research Company."

"Is the Colonel questioning my integrity?"

"I was Battalion XO my first tour and Assistant Division G 4 my second and I certainly would have known of this Radio Research Company. Also, I have studied the Viet Nam war extensively and have never heard of any Radio Research Unit."

"Would the Colonel like to place a wager?"

"Certainly, but I would require proper documentation with you serving in this unit."

"Would you buy my girlfriend and me Surf and Turf next Friday when I provide you with the proper documentation?"

"That is a reasonable wager."

The following Friday at happy hour I presented the Colonel with a copy of my Purple Heart orders from the 22nd Surgical Hospital with rank, name, ASN and Unit, the 265th Radio Research Company (ABN), 101st Airborne Division, APO San Francisco. Case closed!

The COL was very gracious in defeat and we had the club Surf and Turf special that evening with his wife at his expense. Over dinner he told me that he had never heard of the 265th until our meeting.

The following week after our happy hour encounter he said that he had asked several of his Senior NCO's who were in the 101st if they had ever heard of the 265th Radio Research Company. They both had essentially told him we were the Division's signal intelligence company. He spoke with his Brigade S-2 NCO and asked him what he knew about RRU's in Viet Nam. He told him that he had two tours with RRU's in Viet Nam and what do you want to know about them. He gave him the proverbial fifteen minute block of instruction on RRU's mission. After talking with his NCO's he said he knew that he was screwed and was prepared to pay the feed bill.

The White Picket Fence
By Loyd Lavender

This scene will be difficult to describe, because at the time it occurred no one saw it coming and no one wanted to get close to the belligerents. To top it off, it is all true. The unit and place was ASA Group at Pyongtaek, South Korea in about 1972. The Deputy Commander at that time was a Major C who was from Texas. Major C had the shortest fuse of any human being I have ever seen. You could be talking about anything and he would hear something wrong or perceive something wrong and fly into orbit screaming and yelling. You spent most of your time with him trying to calm him down during one of his many tirades. And generally it was because Major C heard something said wrong. If you had a choice you would try to avoid Major C if at all possible.

On the other side of the ledger we had a Korean who everyone affectionately called One Eye. I think One Eye was injured in the war, but I don't know that for sure. One of his eyes was messed up and didn't look like he could see out of it and one of his hands was messed up also. I think One Eye was missing a thumb. One Eye was the handy man and carpenter for ASA at Pyongtaek. I think One Eye moved with the unit from Yongdongpo to Pyongtaek. He was a character. He was funny as hell, especially if you got him drunk, which the ASA NCOs liked to do on a regular basis. He slurred his words and even the Koreans could not understand about half the time. You did not want to get into any long conversations with One Eye.

A third party in all of this was Captain L, who was our Operations Officer. Captain L was the absolute right Operations Officer at this time because he was not afraid to yell straight back at Major C even though Major C outranked him. Captain L was of Italian ancestry and anytime the mess hall scheduled Lasagna, Captain L would walk around Operations telling everyone that they had Italian Soul Food on the menu. One time, the two, Major L and Captain C, were on the telephone; they were yelling so loud they didn't need the device. I don't know why they used the telephone; they were less than a half mile away. Major C hung up on Captain L. Captain L dialed him straight back and said "as I was saying, YOU SON-OF-A-BITCH etc....."

Sometime later Major C convinced Captain L to get a hold of One Eye and have him make a small white picket fence running down the sidewalk between the two buildings in Operations. So he got a hold of One Eye and with a translator explained what Major C wanted. But work got delayed because we had about two or three weeks of solid rain and floods.

When One Eye came to the Operations Compound to build the picket fence, we had tarps out so the pickets wouldn't sit on wet ground while waiting to be inserted.   So One Eye put up the picket fence in very wet soil. One Eye was packing up his tools after putting up the fence when Major C walked into the Operations Compound. One Eye's fence looked like a snake as it weaved down the path. It was crooked to say the least and the pickets were of varying heights.

Major C erupted in one of his usual red-faced, expletive-laced rants, shaking his finger in One Eye's face. Much to our surprise, One Eye did not back down at all. He started screaming at the Major C waving his fist in Major C's face. The two were so loud probably everyone in a 10 square mile area around them could hear their argument.   And obviously, neither man knew what the other was shouting. This shouting match with Major C further enhanced One Eye's status with the ASA NCOs.

The Operations Sergeant was a quiet spoken, nice guy, and he was as nice a guy as you will ever want to meet. He suddenly became the peace maker that day; he got Major C calmed down and we got One Eye out of Operations. Major C, when he left, said One Eye was history.

This scene occurred on a Friday before a holiday, Memorial Day as I recall, that was cloudy and clearing. We had a weekend that was nice and sunny. All of the participants were off that Monday. So when everyone walked into Operations Tuesday morning, the 'day beggars' as they were called by trick workers, were very surprised. One Eye's white picket fence was as straight as an arrow and all the pickets were the same height. The ground had settled and old One Eye knew what he was talking about, whatever the hell that was.

Captain L took one look and immediately ran into the office and picked up the phone and told Major C to come to the Operations Compound immediately. Before this day you would see Major C's face red when he was ranting and raving, but this time his face was red with embarrassment and he countermanded the order that One Eye would never set foot in Operations again. Major C went so far as to call finance and have them increase One Eye's salary. It would have been great if we had camcorders back then to record the battle royal between Major C and One Eye the previous Friday.

An added note: Captain L sent out what I think was the most memorable electric message ever to Arlington Hall Station in December, 1972. It was a time of turmoil in the ASA and NSA because it was the drawdown from Vietnam. Captain L was harassing ASA and, in effect, NSA for not making Pyongtaek an ASA Field Station. Captain L explained that in the past year we were ASA Forward Operating Element Korea (FOEK). He concluded his message by saying, 'and it looks like we are going to be FOEKed again next year.'

Moulapamok
By James Stanton

It was as hot and dry as it can only get in Northeast Thailand in late March. As I walked slowly toward the windowless Operations Building, heat waves rose from the asphalt and it felt spongy under the soles of my jungle boots. I passed through the shadow of the steel-plated guard tower, but it offered no relief. I paused briefly in front of the guard shack so that the bored sentry could glance at the laminated picture ID that was clipped to the breast pocket of my fatigue uniform. He buzzed me through the stout, unmarked steel door. The doorknob was almost too hot to touch. It was 1625 hours, 24 March 1971.

Once inside, the meat-locker frigidity of the air conditioning staggered me. There was a drop of 60 degrees F in ten feet. It was like entering the engine room of some mighty ship. There was the wheeze, clank and rattle of the huge air conditioners on the flat roof just a few feet above my head. The electronic whir of dozens of different machines and the static crackle of the radio receivers added to the noise. Here and there a few low voices were muffled by the background noise. I made my way down a dimly-lit corridor past a half-dozen unlabeled doors. I opened the seventh such door and entered a large low-ceiling room. In a far corner of the room stood a line of four gray standard Army-issue desks shoved together in a row. Behind them were two equally gray metal tables pushed flush against the wall along with a half-dozen metal file cabinets marked 'Expendable' by someone whose spelling ability was a bit shaky. On three of the desks sat beat up typewriters. On one of the tables sat a library of books and periodicals in wildly mismatched sizes that included *Jane's Fighting Aircraft*, a huge phonetic Lao to English dictionary compiled for the CIA, a stack of topographic maps of Laos and Thailand, and a 16-volume official Chinese Communist Party Biography of Mao Tse-Tung translated into English so unreadable that it was impossible to decipher. Inside the file cabinets resided three years' worth of reports, translations and raw radio traffic.

There were no personal touches. There was nothing to indicate what this little cubbyhole of desks and tables was. It had been home for me, however, for almost two years. It may have been Spartan and cheerless, but it was home. I sat down in my chair feeling the well-worn give of its cushion and hearing the familiar creak of its springs as I leaned back. A folder labeled *Day Book 24 March 1971* lay squarely in the middle of my desk. I could see that the folder was fat and heavy. A lot of activity - it promised to be a busy night. I was not surprised. Things had been building up across the river in Laos for weeks now. Nothing big had happened yet, but something surely would before the forerunners of the monsoon rains arrived in another four to six weeks. It was the height of the fighting season in Laos.

I was about to open the folder when Jerry came in. He was our NSA Rep and our effective, though unofficial, fearless leader. If he was still here there must be something up.

"Looks like a lot of skinny" I said, patting the folder.

"Things are really poppin', especially in MR4; CI53 is maintaining continuous contact with HQ in Pakse. They think they'll get hit tonight - a battalion of NVA regulars, maybe two. RJ's got 'em. Roger's at the dentist at the airbase in Udorn. He'll be back by 1800 hours. Frank is staying on base; he's on call if you need him. I'll be on base too, at the officers' billets. Where's Mac?" Jerry asked.

"Where do you think?" I replied, "Drunk, passed out, or on top of some cheap Thai whore in Udorn; those are my best three guesses."

"I've had it with that drunken son-of-a-bitch. If he's not here, and sober, by 1730, he's done. Let him fill sandbags for the rest of his tour. He's no damn use to us. I'll go talk to the young Major. No use talking to these fucking sergeants. They're all drunks themselves - drunks covering for drunks. That's the name of the game in the Army. Well, I'm not in the fuckin' Army anymore so I don't have to take that shit. You'll have a new Traffic Analyst for second shift by tomorrow. I guarantee it. We've got too many bodies on the day shift anyway", Jerry fumed. "Should've canned him long ago." He glanced at his watch. "That's it. That's close enough."

Mac never did come to work that night. He never came to work in Ops again. Jerry and the young Major made sure of that. I read the translations and reports on the top of the pile in the folder to get up to speed. Then I went next door to the Radio Receiving Room to see how RJ was doing. The Radio Receiving Room was bigger than the Translator/Interpreters Room. It had to be to accommodate all the equipment. Racks of radio receivers lined all four walls with a desk, chair and typewriter next to each set. Twenty receivers in the room all told. At 1730 hours, however, 19 receivers stood vacant and only RJ was at his station.

"Where did all your buddies go?" I asked.

"Mess Hall. We're back on 12-hour shifts. Not enough ditty-boppers. They all decided they needed something more than cokes, bad coffee and those repulsive, purple Thai hot dogs, which is all you can get at the Ops snack bar. Besides we have no scheduled eavesdropping 'til 1800 hours. Who needs 'em anyhow? We've got the best Lao lingy and the best ditty bopper radio head in the shop right here. Between the two of us we can run the whole damn show!" he grinned.

"So what are they saying in Laos?" I asked.

"Just chitchat so far. Keeping the frequency open. A lot of "Q" signals. Talk about family. That sort of stuff." RJ replied.

"If you keep doing the Lao net long enough we won't need any lingies. You can do all the translating yourself," I chuckled.

"If they stick to sex and money I'll have it covered thanks to all the language training I'm getting from my Thai girlfriend, but I'm afraid you'll have to help me on the other stuff," he giggled.

"So let's take a look" I said, ripping off a foot-high stack of six-ply paper and leaving RJ with only the page he was currently typing on. I hefted the paper onto the empty desk next to him and began to flip through it page-by-page.

"Who's 4QR7 on 2750 MHZ?" I asked.

"That's BP104 at Ban Ninilchik. He was on gabbing with Pakse and Moulapamok earlier but now he's off," RJ replied.

I lugged the 6-ply traffic back to my desk and tore off the top copy for reference. Roger came in and I briefed him on the situation.

"Where's Mac?" he asked.

"No show" I replied, "and gone forever if Jerry can get the young Major to give him the boot."

"Good riddance!" Roger replied "An empty chair is better than Mac. At least we won't have to take the blame for his continual fuck-ups."

Every evening at 1900 hours military units of the Royal Laotian Army (RLA) began sending their daily activity reports back to their headquarters in the five military regions (MR's) of Laos. There were a lot of translations to get out and send on to the NSA Headquarters at Ft. Meade, Maryland, better known as 'The Puzzle Palace'. Everything had to be scanned quickly and prioritized. The most important stuff got sent out ASAP as a 'Critic' report of whatever was most important with a full translation to follow. Routine stuff was back burnered for the day shift. On the night of 24-25 March 1971, there was a lot of stuff and a higher than normal percentage of it was too important to be back burnered. Roger and I divided Laos in two that night. I took the south and he the north. By 2330 hours all the daily reports were done and all the radios in Laos had shut down except Moulapamok. I was finished with my translations. Roger was still laboring doggedly away on a long and complicated message from MR2. Hmong General Vang Pao was sending details to Vientiane HQ regarding the evacuation of thousands of Hmong refugees from a camp near his HQ at Long Cheng which had been under sporadic NVA bombardment for some time.

"How's it going, Rog? Need help?" I asked.

"Not yet, it's just slow going...a lot of stuff. It took their operator two hours to send it all. It's all in that old fricking French code. I don't know why they bother. Everybody and his brother can read that simple shit. Just takes so long to break it out," he whined. "I want you to review it when I get it done."

I didn't get to help Roger that night. A few minutes later I got an emissary from the Radio Receiving Room.

"Come quick, RJ's got something for you!" he gasped.

So I went as quickly as he did. At 2340 hours on 24 March 1971, the crude little fortress of CI53, Company 3 of the 5th RLA Infantry Battalion with a reported strength of 139 men at Moulapamok in far Southern Laos, came under heavy mortar bombardment. It was the kind of bombardment that usually signaled an impending attack. In this case it was the impending attack of one or more battalions of the 316th Division of the North Vietnamese Communist (NVA) regular army. I rushed back to my desk and grabbed a legal pad and pencil. Then I pulled a chair up next to RJ's desk and began reading the message as he received it in plain un-coded Lao and trying to phrase a 'Critic' report for immediate broadcast. Five minutes later I rushed over to the Com Center which sent all our outgoing messages. I sat down next to a Morse code operator and led him through my message which read:

> *Code ZULU message from CI53 at Moulapamok Laos*
> *Coordinate XA147653 to HQ MR4 at Pakse as follows:*
> *Heavy mortar bombardment commenced 2325 hours.*
> *Enemy now at perimeter wire coming from E and NE.*
> *Heavy ground fire and mortar incoming continues at our*
> *central bunker and entrenchments. Enemy attack now*
> *in progress. Estimate at least one enemy battalion.*

The report was off and winging by 2353 hours, 8 minutes after we got the information and 28 minutes after the bombardment began. That's as good as we can do, I thought. I took my pad and pencil and went back to RJ. I carried three more updates to the Com Center during the next hour or so. At 0117 hours on 25 March 1971, I sat down next to RJ yet again. He typed 'SPOOKY MA SPOOKY MA LEEW' onto his six-ply roll. We grinned at each other. No translation needed for that one.

"The cavalry has arrived, boys!" RJ announced to the room.

'SPOOKY' was an AC-47 gunship out of Ubon Airbase in eastern Thailand. An ancient, lumbering military version of the 1930s Douglas DC-3 fitted out with a high speed mini-gun that spewed out thousands of rounds a minute - a mechanical angel of death. The translation was literally "Spooky's here. Spooky's here already." You could almost hear the elation in the dots and dashes as the Moulapamok operator banged out the message...'Spooky', the angel of death for the NVA if they were caught working their way through the razor wire on the perimeter - the guardian angel for CI53.

Of course we didn't know where the NVA were at the time of Spooky's arrival. We didn't know if they had been caught out in the open. They rarely were caught. Maybe they had withdrawn and would be back later. Maybe at least some of them were already through the wire and hunkered down inside the perimeter waiting for Spooky to leave so they could continue their assault. Only time would tell; time, and perhaps, the coming of the dawn, if CI53 survived that long. At 0205 hours on 25 March 1971, CI53 at Moulapamok went off the air and we thought the worst had happened. So did Pakse HQ as they kept calling and changing frequencies looking for CI53. At 0354 hours CI53 came up again saying that his radio set had been damaged and that he had to shift over to battery power as the generator had been knocked out by the bombardment. I cobbled together a report from his broken chatter.

"His signal's wavering. Still problems with his set, I think" said RJ. "I can get most of it though."

My report: 0410 hours 25 March 1971, message from CI53 at Moulapamok Laos coordinate XA147653 to HQ MR4 at Pakse as follows:

Incoming mortar rounds continue, but light and sporadic. Occasional ground fire from beyond the perimeter. Very dark. Cannot see due to loss of generator power and no remaining parachute flares. We have casualties both KIA and WIA. Enemy casualties unknown. Perimeter wire may be breeched on the east side. Will keep contact if my batteries last. Will call every 30 minutes to conserve power.

I went back to my desk. Roger continued to work on the Hmong Refugee Translation. I thought of the CI53 guys hunkered down in the trenches with their dead and wounded. I thought of the villagers of Moulapamok. I had never been there but I had been enough places in Laos on previous temporary duty assignments to know what it must look like. The gray unpainted wooden houses were in contrast to the bold hues of the village Buddhist temple. Houses scattered along one or two meandering dirt streets that hugged the banks of the Mekong. The villagers spending the night in trenches they had dug themselves.

I had the topographic map including the village spread out on my desk. It was an insignificant village; small, isolated, and of no conceivable strategic importance. Why would the NVA even want to attack it? I wracked my brain for an explanation. Finally, it dawned on me; 'To send a message'. To send a message to the Lao people that it was all over. For the old regime, for the Americans, for anyone who had had anything to do with either of them. The NVA were on the banks of the Mekong. That was the message and they wouldn't need any propaganda broadcasts to send it. The fact that it was not Pakse, or Thakhek or someplace bigger and more important didn't make any difference.

By then it was 0440 hours and I went back to check with RJ. "Nothing" he said. "Just 'Q' signals making sure that the line is still open".

I returned to my desk. I thought about the RLA soldiers waiting, the villagers waiting. Then I thought about Spooky waiting too. Parked out on the runway at Ubon Royal Thai Air Force Base, gassed up, its mini-gun reloaded. The crew in their flight suits probably drinking coffee at this, the only cool hour of the day in March in Thailand. 0510 hours passed. At 0540 I knew that the first pale light of dawn was just visible. It was the time of day when rural Thais get up and cook their day's batch of sticky rice.

At 0540 hours Moulapamok said it was quiet for the first time in 6 hours. Will send out patrols as soon as it is light enough unless we get attacked, Cl53 reported hopefully. At 0610 hours: Patrols out. Will send situation report when they get back. At 0640 hours: No enemy activity. Situation report in 20 minutes." At 0700 hours on 25 March 1971, Moulapamok reported to HQ MR4 in Pakse as follows:

Patrols sent out to check the perimeter have now returned.

174 NVA KIA within one meter of the perimeter wire. More

KIA outside the fence, but number unknown. Fence breeched

in two places...now being repaired. Enemy appears to have

withdrawn. No activity for 2 hours. Our losses 17 KIA and

19 WIA, 14 of those seriously wounded. Need replacement

generator, radio, batteries, parachute flares, and medical

supplies. Ammunition low.

I had just come back from sending the message at the Com Center when Jerry walked in.

"It's all here" I said, tossing a copy of the report I had just sent onto the pile on top of my desk. "The NVA 316th tried to send a message to the people of Laos, but it didn't quite make it... this time. I've had enough, my 16 hours are up, breakfast and then to bed. Hope you don't have any questions, but if you do you know where to find me."

I walked out and dragged myself along the asphalt back toward the residential area of the base. It was 0805 hours on 25 March 1971, and it was already hot. The bullet-riddled bodies of at least 174 dead NVA soldiers were already beginning to bloat in the sun at Moulapamok. I had helped kill them. Maybe I was even the critical factor in their deaths. One hundred and twenty-two RLA soldiers of CI53 were still alive. Perhaps I was responsible for that too.

My First and Last Flight,
Almost
By Ollie White

It was 1979, and after seventeen years in the Air
Force, I was about to start earning my wings as an
Army Aviator. I was a brand new WO-1 and had
just transferred from the Air Force to the Army, and
was on my first flight, on flight status as an Army
Aviator. Two things in life you never want to hear;
the first is your doctor saying "Oops!" when you are
under a local anesthesia, and still wide awake, while
the surgeon is performing a procedure on you. The
second is when you are flying, and can't see what is
going on up front, and you hear the pilot, or co-pilot,
say "What the F--k!" Neither of those two
statements will give you a feeling of confidence in
what is going on around you.

We were flying in one of our two RU-21
aircraft, which was a modified twin engine
Beechcraft King Air -- the aircraft was so small that
everyone in the back end has a window and aisle
seat at the same time. I was on my inaugural flight
after going on flight status; we took off from Biggs
Army Field, at Fort Bliss, Texas, on a training
mission, and headed north east toward the old
Walker AFB, near Roswell, New Mexico. Walker
AFB was officially closed in 1967 and had since
been redeveloped by civil authorities into the
Roswell International Air Center.

The chatter on my headset, between the pilot
and co-pilot, was that the weather report said that
flying conditions over Roswell were horrid; high
winds, rain, thick clouds, and zero visibility, making
it ideal weather for instrument flying. Normally, if at
all possible, most sensible pilots avoid wind, rain,
and clouds like the plague, but our pilots always

went looking for bad weather. Except for 'flying under the hood,' which all of our pilots said they hated because it wasn't as realistic as flying in bad weather, the only time they could get in instrument time was when they could find bad weather some place close by.

I was in the operator's position, on the front right side of the aircraft; it was kind of eerie to look out my window and see nothing but rain pelting the Plexiglas. Outside my window, I could see no horizon, and, except for the right engine, no point of reference; and only my inner ear to tell my brain that the airplane was still flying right-side-up. I don't know what it was like up front, but in the back end of the airplane, I was bouncing around as if I was riding on a bucking bronco. After an hour of beating back and forth in the 'soup' in a race track pattern and practicing 'touch-and-go's', we landed, and the pilot and co-pilot switched places, so both pilots could get in instrument time in the left seat.

Over my headset, I heard the pilot and co-pilot going through the preflight check list; we then rolled down the runway picking up speed. Just after I heard the pilot say "wheels-up," the starboard engine started to stall and the propeller slowed, as if the prop was being 'feathered.' It slowed so much that I could see the individual blades turning, the aircraft yawed to the right, the port side of the aircraft tilted upward, the starboard wing dipped down-ward, back toward the runway, and it's then I heard the pilot yell into the intercom, "What the F--k!" That is not something you want to hear on any flight, but especially your first flight; at that point, I almost peed in my flight suit. In what seemed like a lifetime, but couldn't have been more than a millisecond, because the right wing and propeller didn't hit the ground. The starboard engine roared back to life, the propeller started spinning fast again and we lifted back into the air on a more even keel.

In few minutes, the pilot called me on the intercom to see if I was alright.

"Yes," I replied, in a shaky voice.

"It scared us too," he said.

I guess the excitement dulled their enthusiasm for any more flying that day, because after a couple of laps and a few more touch-and-go's we headed back to Fort Bliss. After we landed, I learned from the pilot that the friction knob on the starboard throttle -- the knobs that are supposed to hold the throttle in the full power position during takeoff -- wasn't cranked down tight enough so as we lifted off and the aircraft started to climb, the starboard throttle fell back into neutral position and the engine, starved of fuel, started to stall. As soon as the plane yawed to the right the co-pilot looked down at the throttles, saw what had happened, and rammed the throttle full forward and we picked up speed again before we lost anything critical to future flight -- like a wing tip!

A Terrorist Is in
The Eyes of The Beholder
By Ollie White

It was 1979, and I was assigned to the American Embassy in Belgrade, Yugoslavia, as a Signal's Intelligence Analysis. While the threat of terrorist attacks hadn't yet crossed the Atlantic Ocean to America, terrorist attacks, or the threat of terrorist attacks, were a way of life in Europe. We had been at the embassy about six weeks, when I finally got notification that our car had arrived at the port of Split. I wasn't allowed to travel to the coast, so I made arrangements to pay one of the Yugoslavian drivers in the Attaché office to go and pick it up for me. It was just as well, as I didn't have a way to get to Split, except by bus or have someone take me, and since I didn't speak the language, it would have been very difficult to handle all of the customs paperwork at the port. The car arrived in good shape, and I had the only Subaru -- Serbian pronunciation Zoo-baa-roo -- in Belgrade and most likely all of Yugoslavia.

I had the embassy send a request to the protocol office so that I could get my diplomatic license plates, but in the meantime, I was allowed to drive around town with my Texas tags on the car. The American Embassy was in the middle of old downtown Belgrade and we had no recreational facilities. However, out at the ambassador's residence there was a swimming pool, and Ambassador Eagleburger was gracious enough to allow embassy employees to use the pool. His only stipulation was that if we came out to the residence, we not use the front gate but go around back and use the rear entrance. The ambassador's residence was located 7-8 kilometers south of downtown Belgrade in the Dedinje district in Belgrade's municipality of Savski Venac. The Dedinje neighborhood was generally considered the wealthiest part of Belgrade and was the site of numerous villas and mansions owned by the members of the city's wealthy plutocracy, as well as many embassies, diplomatic residences, and Tito's residence.

Following the end of WW II, the new communist government of Yugoslavia seized the assets and property of the royal family, including the White Palace in the Dedinje district of Belgrade; the White Palace was Tito's residence when he was in the capitol.

The first weekend after we got the car, the kids wanted to go swimming, so on a Saturday morning I grabbed the map and we started out for the ambassador's residence. I had no problem getting to the Dedinje neighborhood or finding the main boulevard with all of the upscale restaurants, villas, mansions, embassies and diplomatic residences; all I had to do was follow the smell of money. It didn't take long to find the American residence compound. Since my wife Linda and I had been there a couple of times at receptions given by Ambassador Eagleburger, we had always come in through the front gate. I drove on by and continued down the boulevard until I got to a cross street where I could turn right. Unlike the main boulevard that was straight, the side street did some twists, and I had to drive several blocks before I could make another right turn, which should have taken me onto a street that ran back behind the ambassadors house, but, again, there were a few more twists and turns, and when I got back to where I thought the residence should be, nothing looked the same or familiar. I continued on the road behind the residence for a about a mile, passing numerous large estates and houses, until I could turn right

again and make my way back to the main
boulevard. We made another right turn, completing
the loop, and started back toward the ambassador's
residence.

When we got in front of the residence again, I
stopped and tried to get my bearings by focusing on
the landmarks at the rear of the house, but with a
wall around the grounds, and the house in the
center of the property, it was hard to see anything
that would give me a point of reference. We
continued to the side street, made a couple of right
turns, and again I found myself on the street that
should be directly behind the residence, but all I
could see were walls and trees. I paused a few
times to look around, then made my way back to the
main boulevard and made the loop again.

I had forgotten an important lesson from my
previous tour in Germany: in Europe, you should
never try to go around the block. This time when I
got back around to the back street, all hell broke
loose. The strange, blue, foreign car, with the
unknown, white, blue and red license plates, which
wasn't on file, had been around the block three
times and now had aroused suspicion.

The guy I had seen walking his dog wasn't
out exercising himself, nor the dog, this morning;
the guy on the ladder with a can of paint and paint
brush wasn't painting houses this morning; the guy
working in the yard wasn't a gardener or hoeing
weeds. These guys were hanging around the
neighborhood to make sure that terrorists stayed
out of the Dedinje neighborhood. The three of
them, with guns drawn, were onto us like 'white-on-
rice.' The gardener came across the street, with a
machine gun, which he had hidden in his
wheelbarrow. The dog on the leash wasn't a
friendly pet; it was now a snarling beast. With
something like panic in my voice, I kept yelling,
"*Američki, Američki*" (American, American). One of

the guys, who seemed to be in charge, was yelling back at me in Serbo-Croatian, none of which I understood a word of, except the words *Lične Karte* (National Identification Card). Everyone, in Yugoslavia, resident, guest worker, or embassy emissary, was issued a *Lične Karte* and was supposed to carry it with them at all times. I didn't have my diplomatic passport with me, but did have the *Lične Karte*, so produced it. That seemed to calm everyone down; the guns were put away, and one of the guys, with a walkie-talkie, spoke into his radio. In about fifteen minutes, an English-speaking official drove up. He wanted to know why I kept driving around the block. I explained to him that I was looking for the back entrance to the residence of the American ambassador and that we were invited out but were told not to use the front gate, but every time I came around back I seemed to get lost; that was why I had been around the block so many times. He spoke to the other three guys and they went back to doing whatever they were doing before.

Ollie White, the suspected terrorist, had been around the block three times. The English-speaking official told me to make a U-turn and follow him, and he would show me where the back entrance to the residence was. A couple of blocks back up the street, we turned into a small alleyway, which was almost completely hidden by trees and behind the trees was a small parking area, the back wall to the residence, and a small gate. The road was so obscured, that if he had not shown me, I don't think I would have found the back entrance in a million years.

Chapter 6
Reflections on Vietnam

Do You Ever Dream About Vietnam?
By George Murphy

I don't; well hardly ever and almost never. However, I do think about it often.

I was at the 3rd RRU during part of 1964 and 1965. Back in the States I went to Ft. Devens for training and then to Ft. Bragg. I was at Bragg when the 313th ASA Battalion received its orders to deploy to Vietnam. At that time, the Army could not just send you back to Vietnam. My options were deploy to Vietnam or go to the Dominican Republic. I volunteered to return to Vietnam.

I was in the advance party. There were eight of us. We arrived in Saigon in January 1966. We were sent to the 8th RRU at Phu Bai for training and orientation. The 8th was up near the DMZ. When the 313th, now called a Radio Research Battalion, arrived in country, we moved down to Camp McDermott just outside of the seaside town of Nha Trang. It was delightful tent city near the South China Sea. After a while at Nha Trang, I moved over to Det 3 of the 3rd RRU. It later became the 406th Radio Research Detachment. It provided SIGINT support for the 1st Brigade of the 101st Airborne.

While at Det 3, in addition to the standard Morse code intercept activity, I was involved in some low level voice intercept operations. We would go out in the field with the infantry units on their combat operations. If I remember correctly, I was mostly with elements of the 2nd battalion, 502nd Infantry. Its nick name was Gunfighter. When I did low level voice, I had two ARVN's with me. They monitored the radio looking for VC/NVA activity. In my tenure, we almost always heard ARVN communications and no VC/NVA. I understand that other units had a different experience.

Fast forward forty plus years; my wife and I are driving from our home in Ohio to visit my brother-in-law and his family in Florida. Our trip takes us on I-75 out of Ohio. We pick up I-40 east of Knoxville, and then I-26 in the Carolinas and then on to I-95. By the time we reach I-95, we have been traveling for about ten hours. We called it quits.

We stop at a motel just off the interstate. It is an older facility. It has outside corridors. The room has the AC/heater under the front window and next to the door. The unit is loud and it runs continuously.

After some dinner and TV, we go to bed. This is where my dream begins.

I am in a jungle clearing. I am told that choppers are on their way to take us out. I must be doing low level voice, because I have two ARVN's that appear to belong to me. There are also about thirty grunts from the 101st. It is starting to get dark. We are all hunkered down in the trees at the edge of the Landing Zone. Things are quiet. Everyone seems relaxed. There are a lot of people smoking.

The man in charge is a Captain. He is airborne with a CIB. My guess is West Point. He has that air of authority about him. His radio man is attached at his hip. The quiet is broken. The radio barks.

"Red Ryder, Red Ryder, this is Bright Star Leader, ETA your LZ five minutes."

The Captain takes the horn, "Bright Star this is Red Ryder. We are sitting here looking for a ride home".

"Roger, Roger, Red Ryder, we are on the way."

Things go quiet again. The Captain is pacing the tree line. All of a sudden the Captain is back on the radio. "Bright Star, Bright Star, this is Red Ryder, we have a situation." The grunts are beginning to get nervous.

"Red Ryder explain."

"Bright Star there is something on fire at the far edge of the LZ and we do not know what it is."

The radio goes quiet. After a few minutes, Bright Star is back.

"Red Ryder stay put. We have a gun ship flying cover. They will investigate. You stay. We are coming in. Throw your smoke. Throw your smoke."

Just then a guy from the 101st jogs out into the LZ and pops a smoke grenade. Yellow smoke billows up from the LZ.

The radio is back. "Red Ryder I see your smoke. Your smoke is yellow."

"Roger, Roger, yellow smoke. Get us the fuck out of here."

Just then you could hear the whoop, whoop, whoop as the Huey's come in low over the trees. They settle down in the clearing. A sergeant from the 101st motions for me and my ARVN's to board a chopper. We jump on board. The Huey lifts off. In the distance, I can see something on fire, but I cannot make it out.

My dream ends.

The next morning, I tell my wife that I had this vivid dream about Vietnam. Honestly, I really do not have Vietnam dreams. We make our way to our complimentary continental breakfast. During the meal, a cable TV station reports that I-95 in South Carolina was closed due to an accident. Something was on fire. I say crap. Seven hours to go in the car and I do not need the highway closed.

After eating, I go to the lobby to check out. I ask the clerk if she knows if the closure is north or south of here. She says, "It is not north or south of here. The accident happened out our front door.

You can see the burned out wrecks out our front window. You slept through that? We had police cars and fire engines in our parking lot most of the night with their radios blasting. They even landed helicopters out front to evacuate the injured. It seems most of our customers were out in the parking lot in their PJ's watching the whole thing. I cannot believe that you slept through it."

I said, "I think the AC unit was noisy and at another time in my life, I heard lots of helicopters taking off and landing." After check out, we hopped into our car and headed south to Florida. That was the last time I dreamt about Vietnam.

Do You Want a Python
With Your Coffee?
By George Murphy

I am retired. In the winter my wife and I go to Florida in February for four, five, or six weeks depending upon the situation. We go to the East coast, Jensen Beach. We rent a condo in the same development as my brother-in-law and his family. They live there year-round.

Every morning I would walk across A1A to a Cumberland Farms and buy a newspaper. I would have my coffee and read the newspaper on the front porch of the condo.

One morning, I walked across the street, and there was a guy holding the door open to the Cumberland Farms' store. He had on a hat that said Special Forces - Vietnam. I made a comment that he was doing a good job in keeping the riff-raff out and by the way I was in Vietnam myself. We exchanged thumbs up.

A few minutes later we were together in the checkout line. He asked who I was with in Vietnam. I said the Army Security Agency - ASA. He never heard of it. I said that is not surprising. It was a relatively small intelligence unit and its existence in Vietnam was classified.

The conversation continued. "Where were you?" I explained that I was at Tan Son Nhut, Phu Bai, Hue, Nha Trang, Phan Rang, Thuy Hoa, and Dak To.

He told me that Special Forces had an A Team at Dak To and that he was there about six months after I was. I explained that I was at Dak To with the 1st Brigade of the 101st Airborne.

He asked me if I had ever been at the Special Forces camp. I said I had been there and here is what I remember. A shower point had been set up at the camp. They pumped water out of a stream into a large rubber tank. You could take a shower and get a clean uniform. The shower was in a tent. The floor of the shower was a collection of wooden shipping pallets. The pallets were the kind that a fork lift would use. Wooden slats on the top and bottom with several inches of open space in the middle.

I told him that while I was taking a shower, there was a python crawling in the space between the top and bottom of the pallets. Being ASA, I headed out. The Special Forces types went to work to capture the python. Sometime later I returned to the camp. I do not remember why I was back there, but the python was in a cage.

The Special Forces guy said, "Holy Crap. I went out this morning to get a cup of coffee and I meet a guy who remembers that the A Team at Dak To had a python as a pet 46 years ago. Who would have ever guessed that?"

I said, "Sometimes it is a small world."

America Isn't Free
By Jerry Frankenburger

There was an old Vietnamese woman who decided to go to the U.S. to visit relatives that had emigrated there during the aftermath of the war. She had a sister that was living in Pittsburgh, Pa., and made her mind up that she was going to see her. The old woman had saved up quite a bit of money during her life and decided to use some of it to help her sister buy a home in Pittsburgh. The situation being that most Vietnamese use gold for their savings she was going to take some of the gold with her to give her sister. Her son explained to her that she could not take gold from the country, but she should turn her gold into a cashier's check. She would have no part of this thinking, that somehow a piece of paper could be worth the gold. After family members explained that she could take cash, she went that route. But she didn't want to carry it in her purse, believing that someone might rob her. So she had a daughter-in-law make her a double blouse with pockets between the layers; that way she felt that she would be safe.

She got her ticket to Pittsburgh by way of Hong Kong, San Francisco and Pittsburgh. Her family saw her off on her flight and the beginning of her trip was A-Okay. At her first stop in Hong Kong, being the only language she spoke was Vietnamese, she could not communicate, so she decided to follow other passengers from her plane. As the other passengers walked to their respected destinations there were fewer and fewer for her to follow. She would not seek help, being she had a very strong distrust and dislike of Chinese. She ended up missing her flight.

A week went by and there was a knock at her sister's door; there was a TSA agent delivering her to her sister's home. The agent said keep a close eye on her and explained that her trip to Pittsburgh had taken a roundabout route. She had gone to Bombay, from there to Germany, then to New York and finally Pittsburgh.

While visiting she decided to help out around the house and reasoned she could make Pho (Vietnamese Noodle soup) and sell it. She got everything ready and went out front of her sister's home to sell her soup. Her sister caught her before the health department or police and made her get back inside.

There were some young kids in the neighborhood that marked her as a target for some verbal abuse. She went to their parents and complained, but the parents would not do anything with their kids. She took it upon herself to do something about it. The next time the kids started harassing her she chased one into his home, right past his mother, caught him in the kitchen grabbed a big spoon and gave him a licking. The mother was so scared she could not react. She did call 911 and had the old woman arrested. She ended up in court and had to pay a large fine. On the way back to her sister's home they ran into a traffic jam. She said to her sister just drive on the sidewalk, it was clear. Her sister explained to her that this could not be done in the States.

She went home to Vietnam soon after this and explained to her family that America was not free. They asked why she thought this and she explained you can't sell Pho on the sidewalks, you can't drive on the sidewalks and you can't punish undisciplined kids. This story was told to me by one of our guides about his mother on my last trip to Vietnam.

The Medal
By Skip Saurman

Late afternoon on a stinkin' hot and muggy Sunday, August 22nd, in 1971, I was standing in a rather small and somewhat loose formation with Sergeant First Class Robert L. King and Sergeant First Class Edward D. Petersen in the middle of a dilapidated concrete basketball court between the barracks near the center of Camp John F. McDermott, a U.S. Army Base located just outside the coastal city of Nha Trang, South Vietnam. I was dressed in old, dirty, wrinkled, worn-out, un-bloused army green fatigues and a floppy boonie hat with the sides bent way down over my ears. It was tilted forward with a strap that ran tight around the back of my head to hold it on. So, there I was, standing in what some may refer to as attention while the 330th Radio Research Company Commander pinned a brand new, shiny Bronze Star on my left chest pocket.

I managed a slight smile and gave him a half-hearted salute, but I was weary from approximately ten months in a war zone, so afterwards I took it back to my hooch and packed it up in order to ship it home. After all, what good is this particular hunk of metal in this God-forsaken place? All I' was really thinking about those days was getting me home in one piece. You see, as you serve longer and longer in-country, and your time becomes shorter and shorter, all you can think about is going home to a hot private bath because all you can ever get here is a makeshift community shower and some maybe edible food, but all you're ever thinking about is a real movie in a real movie theater, fast cars, round-eyed women, and so on, you know, the stuff in the *real world*.

We were all well aware of the stories back in the States about the lack of support, to put it mildly, regarding the war in Vietnam. Reports about being ignored, cursed at, spit upon, yelled at, labeled War Mongers, Baby Killers, you name it. Well, during that time we had a mailman back home by the name of Archie Montgomery, who was just an everyday, ordinary, average mailman except for one little thing. Every single time that there was a letter or a package home from me in Vietnam, Archie would always bring that letter or package straight up to the front door, ring the doorbell, and personally hand-deliver it to my Mom. He just knew how important that was to her! While I was in the 'Nam, I would write letters home pretty much religiously, every week. However, serving in a Radio Research Unit, actually, the Army Security Agency, but everyone knows that the ASA really wasn't there, I could never officially discuss exactly what I was doing over there. Still, whenever Mom didn't receive a letter that week, she just knew that something was going on.

Now I'm pretty sure that Mom really didn't know what a Bronze Star was, but she must have figured out that it was something kind of special. Or maybe she just asked around.

For the next thirty-two years, Mom would continually ask me "What was the Bronze Star Medal for, Skip?" "What did you do to get the Bronze Star Medal, Skip?" "Why did you get a Bronze Star Medal, Skip?"

Well, when you first return from a place like Vietnam, all you want to do is fit back into the world and forget about all of the unpleasant memories experienced in a war zone. I remember that it was roughly ten years before I would even watch any kind of war movie like The Deer Hunter and even much longer than that before I was anywhere near comfortable enough to talk about any of my experiences and escapades. So, my standard answer to my Mom's consistent questioning became, "Aw, I was just standing in a line and they came along and pinned it on me."

She never did accept that flimsy explanation, even though it seemed more or less true to me, and until the day she passed from this earth, her desire for the truth continued relentlessly.

If I were to have a regret in my life, it is that she went to her grave without the satisfaction of knowing why her son received a Bronze Star Medal while serving in the Republic of South Vietnam! It wasn't like I didn't know that the end was too quickly approaching for her. The very last time that I was fortunate enough to talk to her was at the side of her hospital bed in Belen, New Mexico, in early 2003. She was tired and her 82 year old body was worn out from the drastic effects that resulted from a lifetime of cigarette smoking and now, emphysema and lung cancer.

She expressed severe anxiousness and a fear of passing on to an unknown place. All I could do was try and relate to her the anxiousness that I felt as a teenager in 1970 at the U.S. Army Overseas Replacement Station in Oakland, California, where I waited for a bus to take me to Travis Air Force Base and the awaiting great silver bird that would transport me to the faraway paradise known as Vietnam, a 24 hour flight with stops in Anchorage, Alaska, and Yokota, Japan. I also tried to relate to her fears as I conveyed the experience of our landing at Bien Hoa Airfield and was told to keep our heads down while running to the nearest bunkers if we should come under attack while deplaning. Even though that didn't happen that time, I distinctly remember sitting on the barracks steps most of that night, talking and listening to a mortar barrage somewhere off in the distance and wondering exactly what dangers the next 365 days would have in store for me.

Mom didn't ask me about the Bronze Star this time, and I didn't even think to bring it up. All I could say was, "It's alright Mom, I love you" and gave her a great big hug and a kiss. A few weeks later, she was gone.

Following is a long overdue letter that I would have, no, should have written to her, many, many years ago.

Dear Mom,

Please, please, please, forgive me for not writing this letter to you much sooner. I've owed you a decent explanation about how and why I received the Bronze Star Medal, so here it is.

Many people are unaware that there are actually two different categories of the Bronze Star Medal. The Bronze Star Medal is awarded to individuals who, while serving in the United States Armed Forces in a combat theater, distinguish themselves by heroism, outstanding achievement, or by meritorious service not involving aerial flight.

Well, even though I was an adventurous, testosterone driven, rough-and-tumble, adrenaline-rich, eighteen year old when I enlisted in Uncle Sam's Home for Lost Boys, I have to say that I did not perform anything dangerously heroic or dramatically patriotic in order to get a Bronze Star with Valor. No Mom, mine was for meritorious service. Webster defines meritorious as, possessing merit; deserving of reward or honor; worthy of recompense; valuable.

Let me take you back to the beginning. If you remember Mom, I volunteered to enter the U.S. Army during my senior year in high school in order to become a combat engineer, thanks to your brother Eddie's positive influence in my life. It was to be a three-year hitch, but after scoring considerably high marks on the military aptitude tests, I was approached by and then volunteered to join the U.S. Army Security Agency, thereby adding another year to my military commitment.

Mom, remember when I achieved the freedom and independence of a sixteen year old with a driver's license and my *own* car? At least it was one that I could drive as long as I could manage to pay for the insurance, gas, tires, and the upkeep on it. You and Dad would be sitting on the front porch every night as I backed my dark green '56 Chevy Bel Air out of our driveway. When I stopped to say goodnight, because I just knew that you'd both be in bed whenever I returned from my nightly excursions, you would always ask, "Where are you going tonight, Skip?"

My reply was always the same, "Looking for the action, Mom." Well, it seems like I always returned long before your plans for bedtime.

When you asked, "Well, did you find the action, Skip?" my answer was always, "Nah, there's nothing to do in this lousy town." Now, this was Northeast Philadelphia where in reality there was always plenty to do. The problem was that everything would cost me money; money that I didn't have, even for gas.

Anyway, I joined the Army Security Agency looking for the action. I thought I knew a little something about the Agency, since my old tech school automotive instructor had served in that branch. And after all, the recruiter did promise that I would be wearing civilian clothes with a trench coat and carrying a hidden dagger, in some clandestine location. So I worked hard and excelled in all of my classes, becoming a 05D20 -Special Identification Techniques Operator or Duffy and then a 05D30 - Special Identification Techniques Analyst. I did everything that I was told – usually - and everything that I was supposed to do - most of the time. I achieved rank pretty quickly too, going from a private E-1 in the delayed enlistment program, to a private E-2 entering Basic Training at Fort Dix, NJ. I then entered Advanced Training at Fort Devens, MA, as a Private First Class E-3 and went to 'Nam as a Specialist 4th Class E-4, coming out of all this in a measly three years as a Specialist 5th Class, E-5 or similar to a Sergeant in the regular army. I got an early-out of the Army for being in a combat zone.

Here's the funny part about all this, Mom. The ASA provided me with a Top Secret Crypto Security Clearance, the highest security clearance available to anyone in that particular time period, and spent a lot of time, money, and energy to train me for a specific job or series of jobs; Morse code, radio signal intercept, radio direction finding, map reading, code breaking, signal analysis. And, just as I was dutifully assigned, I performed all of those tasks, that is, after first performing about three weeks of various work details I was assigned in Vietnam. Yep, I figured that I actually worked in what I was trained to do for approximately three months. You see, in radio direction finding it is imperative to have as many active sites as possible in order to intercept and triangulate a particular radio signal. At the time, in Vietnam, I think we had about fourteen such sites, along with portable radio direction finding equipment known as PRD Teams that were infrequently utilized.

All of those sites had to be first established with hooch's, huts, and antenna fields; then manned, supplied and maintained. When they found out that I possessed a mechanical background of sorts they said, "You need to be on convoy, boy" in order to effectuate the above-mentioned tasks. Not really a difficult decision for me to make since I never really considered myself a desk jockey. So, I found myself traveling around the countryside, going from site to site, utilizing Nha

Trang as a base camp. Nha Trang also housed the motor pool and maintenance operations. Ban Me Thuot, Bien Hoa, Cam Rahn Bay, Dong Ba Thin, Qui Nhon, Tuy Hoa, Phu Cat, and Pleiku, were a few sites that you may have heard of from the TV news channels.

The vehicle of choice for these missions was the M35A2. It was a ten-wheeled, 6x6, two-and-a-half ton, all-purpose vehicle lovingly known as a deuce-and-a-half. It was powered by a large 7.8L six cylinder diesel engine that was rumored to run on any type of fuel, be it diesel fuel, jet fuel, gasoline, alcohol, or human pee, probably only after you've drunk enough alcohol. We were oftentimes accompanied by a gun truck or two which were also deuce-and-a-halves but outfitted with .50 caliber machine guns, sometimes in a quad mount turret configuration, and sporting plenty of armor. Through the treacherous mountain passes, where travel was oftentimes slow going, we were usually also supported by Cobra Helicopter Gunships, if the convoy was large enough. Once you were away from the relatively, semi-secure feeling of a base camp, you succumbed to the queasy sensation of being out on your own. Yep Mom, here was the action that I thought I was looking for.

But wait, you wanted to know why I got the Bronze Star, right Mom? Well, after quite a few trips, I progressed from being just a deuce-and-a-half driver to driving point in the lead Jeep. Many times, this Jeep also had a .50 caliber machine gun mounted in the back. If someone was going to get hit driving through a mountain pass, it was going to be the very first vehicle in line so as to block all the rest of the convoy and then pick them off one-by-one.

Because of its somewhat central location in the Central Highlands of South Vietnam, Pleiku –

which was Detachment 8 of the 330th Radio Research Company – was a particularly active site and we made many, many trips through An Khe and other mountain passes to provide them with much needed support.

Anyway, the fact that Pleiku was such an active site, it meant many all-day-long supply trips of 400 km or about 250 miles up the mountainous coastline traveling Vietnam's route QL1, a major but narrow two-lane, treacherous paved highway, with no stripes of any kind and no guardrails. We would drive through Tuy Hoa where the bridge was almost always out and turned west onto route QL19 at Qui Nhon, a much worse road than QL1.

An Khe Pass was the first major mountain obstacle and Mang Yang Pass came up shortly after that. These were high, steep mountain passes defoliated with Agent Orange in a somewhat futile attempt to deter the enemy places to hide near the road. Whenever we were hit, it was mostly mortar fire or Rocket Propelled Grenades from a distance. The drill was to grab your rifle and helmet from the seat next to you - the steel pot was way too heavy to wear on your head in the bouncy environment of the deuce-and-a-half - and your flak vest from the cab floor that would hopefully offer you some protection from land mines, as you exited the vehicle and headed for the nearest ditch along the side of the road. From there you would shoot. Shoot at anything. Shoot at everything. If we ever did hit something it was just pure dumb luck. I never saw anything that I shot at.

It was on one of these expeditions that a problem occurred. We arrived at the base camp in Pleiku well enough at the end of a long days drive, but when we were getting ready for the return trip to Nha Trang the next morning, it was discovered that the air brakes on a deuce-and-a-half were inoperative.

You've probably heard all of the stories about government inefficiency of leaving equipment behind, pushing helicopters off aircraft carriers, etc. This was the summer of 1971. Vietnamization and military cutbacks were starting and getting any new equipment to Vietnam was next to impossible. The 330th RRC only had a limited number of vehicles to start with and leaving it in Pleiku would be of no use, since they had no resources available for the repair of vehicles at that time. So, for whatever reason, or maybe a lot of reasons, this truck just had to come back home. But who in their right mind would be foolish enough to drive a two-and-a-half ton truck with NO air brakes through all those treacherous mountain passes? Well, I'm really not usually the one to volunteer for things, but after all, I am the car guy, so, I agreed to drive the truck.

So, it would be a minimal vehicle convoy that hopefully would not attract attention from the VC. I would be positioned in the front of the convoy in order to prevent any catastrophic results in the event of a runaway. I had them empty the truck of any unnecessary weight and teamed up with another driver to run ahead of me just in case I should need to bump him in order to slow down. We checked and double checked hand-held radio communications and we were off early the next morning. We later discovered that the hand-held radios were useless and who the hell had time to talk on the radio, anyway?

It started off good enough. The deuce-and-a-half had a 5 speed manual transmission with a separate dual splitter allowing the use of ten gears to help in slowing down. When we came to the top of the first mountain pass at Mang Yang, I didn't know exactly what to expect, but was prepared for the worst. It was truly a lot of quick downshifting, but a great sigh of relief when I finally made it down to the bottom of the mountain! But, almost right away, it's up to the top of An Khe Pass and then the next pass, and the next pass.

I was feeling pretty good then, almost cocky and getting into the coastal mountain region where the mountains are not quite as steep. But it was here that I can explicitly remember to this day, some really scary, tight curves in the road where I thought I might lose it and almost did!

Well, the good lord smiled on us that day, Mom. And the angel that Uncle Ed loaned me to rest on my shoulder when I first left for this life's oversea adventure, worked overtime. We rolled into Nha Trang late in the evening tired, hungry, beat up - the truck, that is, and me. Charlie was kind enough to leave us alone on this particular trip. The truck was in the motor pool for repairs and we were none the worse for wear after some chow and a good night's sleep. Mission accomplished.

I didn't think any more about it until sometime later a very dear friend of mine, Sgt. Rock, aka Terry L. Shrock from Kokomo, Indiana, said that he would recommend me for a Bronze Star.

"Yeah, yeah, sure Sarge, maybe it will happen, maybe not. We'll see."

Well, a couple of months later, I was standing in formation in the center of that concrete basketball court in the middle of Camp John F. McDermott in Nha Trang, with the Company Commander standing in front of me, pinning this Bronze Star on my chest. It's the one that you received in the mail.

Just about anyone you ever talk to that has received a prestigious medal, including the Medal of Honor, our nation's highest award, will tell you that it was no big deal. "Just doing my job." I once heard about a Medal of Honor recipient that claimed his Medal of Honor wasn't even as important as a fellow soldier's Good Conduct Medal! His medal was earned for a particular action that he had performed one time, while the Good Conduct Medal requires exemplary conduct, efficiency and fidelity during three years of active enlisted service with the U.S. Army.

We all just did our jobs Mom, whatever that may have been. Some received these types of tokens, while many others did not. It was just a matter of being in a certain place or situation at just the right moment in time, and for someone else to recognize it.

A common phrase in the 'Nam was, 'It don't mean nothin'.' The phrase usually was expressed when things weren't going quite right and in reality, I think it really meant an importance that you just didn't want to admit. So, at the time, the Bronze Star didn't mean much to me.

John F. Kennedy once said, "A man does what he must in spite of personal consequences, in spite of obstacles and dangers and pressures, and that is the basis for all human morality."

Well, times change, Mom. It's now 42 years later and I proudly wear a simple Bronze Star ribbon sewn on the left side of my baseball cap. No medal, no flair, no explanation, no description; just a plain, simple, single ribbon. It is there to constantly remind me of my brothers and sisters who have in the past, and those who continue in the future, to defend the freedoms that we cherish here in the United States of America!

So finally, that is my story, Mom. Not particularly worthy of a TV movie like you might have imagined. I am *truly* sorry that it took this long to come out of me, but through writing this letter to you, I feel as though you can now rest in peace, knowing what happened. And, it can now be passed along to Julie and Neal, as well as your grandkids so they won't have to endlessly wonder, like you did. God bless you, Mom.

All my love,
Skip

The Russian Embassy
By Wayne I. Munkel

As I approached my return to the U. S. from
Vietnam in the summer of 1964, I ordered a new
hunting bow and two dozen arrows. I was planning
to hunt black bear in Canada in September. I
returned to my home in Iowa in mid-August. For
about six hours a day I practiced shooting the new
bow and my old one. I was going to hunt with an old
deer hunting buddy from my high school days. By
September first we were in Ontario, Canada. The
first night we were camped in our 1950's panel truck
by a lake. I had just gotten to sleep when I heard
wolves across the lake howling at each other. It was
the first time I had heard wolves and it was a scary
sound.

    During the next week I managed to stick an
arrow in a big black bear but didn't kill him. The next
year we returned to the same area and I found out a
rifle hunter had killed the bear two weeks later. How
did they know it was the bear I shot? When they
dressed the bear it had an X mark from my
arrowhead right behind the diaphragm.

    For the next fourteen months I worked in
building houses and barns in Iowa and Southern
Minnesota. I had worked in construction before
going into the Army. During this time of working
construction, I had decided that I did not want to be
a carpenter for the rest of my life. Maybe it was
nearly getting electrocuted when the power saw
shorted out in my hand. Maybe it was the day I was
pounding nails in minus 20 degree weather.

    I had gone to college for a year before I went
into the Army. ROTC was required back then. The
college experience helped me get into the ASA and

ROTC made basic training a breeze for me. After I was discharged from the Army, I decided to go back to college. I wasn't sure what I wanted to study but I knew the answer was in my going back to school.

I returned to Iowa State University in December, 1965. I tried the track that would take me into the U.S. Foreign Service. Vietnam had given me a wider perspective on the world and I thought I would like foreign travel and adventure. After I did poorly in French, it was clear that I was not going to make it into the Foreign Service.

Next I decided to go with an engineering curriculum. My family was into building construction and I had some experience with them in building houses and farm buildings. I could contribute to the family business. Failing Differential Calculus stopped that field rather quickly.

Along the way I found that I liked history. History involves a lot of government and culture among other things. I minored in government. There were a number of University requirements I had to take and in the process found that I liked and excelled in zoology courses. I graduated in 1969 with a B.S. degree in History and minors in Government and Zoology.

In the seventy's, I did what most twenty year olds do. Get a career going, get married, have children and begin the American middle class life. I worked for the City of St. Louis with delinquent boys from the city for almost ten years and during that time earned a Master's degree.

Along about 1980 I began working at Cardinal Glennon Children's Hospital. This job was to be my life's career. It was there that my earlier education in Government and Zoology were put to use. My background in zoology was perfect for understanding the medical field and the human body. The Master's in Social Work I had earned provided the knowledge necessary to help children and their families.

I have been described as a wannabe doctor and wannabe cop. During the entire time I worked at the hospital, I worked closely with law enforcement on child homicides and child maltreatment. Being able to communicate with law enforcement, child protection workers, and parents was essential to my job. In many ways I saw myself as a translator between the medical staff and the people who needed to understand what the child's condition was in terms they could understand.

Much of my career was spent finding resources to meet the needs of the people in need of those resources. It could be as simple as obtaining medicine or as complex as creating a community wide system for dealing with sexual abuse. Part way through my career I added administrative duties to the list of things I did.

I worked twenty years as a part-time lobbyist advocating for better laws to protect children and to make life safer for them. This was not my primary responsibility but it became a passion. In working with legislators I influenced a number of pieces of legislation that became law. I got to know the process at the state level. I got to know legislators on a personal level. Most were ordinary people who got elected to office. I had not thought much about being in office myself, but knew that I was ordinary like them. But when a city council seat became vacant in my city, my wife challenged me to stop complaining about city government and to run for that office.

I filed the necessary paper work and got on with campaigning in earnest. I wore out a pair of leather shoes walking the city, knocking on doors and talking with people. When the votes were counted, I had won. I was now Council Member Munkel, 2nd Ward, University City. My big head lasted until the first council meeting because it was time to go to work. I won a second election and spent eight years on the Council.

One of the things that happened in our city government was that you became a member of different organizations related to government just by being elected. I found out I was a member of the County Municipal League, the State Municipal League and the National League of Cities. Not only that, but I was expected to serve on committees and attend meetings and conferences related to these organizations. These meetings were in addition to the community meetings and Council meetings that constantly demanded my attention.

One committee that I was selected for, and that I wanted to be on, was the Information Technology and Communications Policy Committee of the National League of Cities. What else could you expect from a former ASA radio operator? The committee made recommendations for or against the FCC actions at the time. It was interesting work and I enjoyed it a lot. I even testified before the committee a couple of times.

The Congressional Cities Conference of the National League of Cities was held each year in Washington, D.C. I lobbied on Capitol Hill during the conference and was able to meet many of our Congressional leaders over the years. I learned that many cities across the country were struggling with similar problems and issues as we were. While my city was in the forefront of many issues in the St. Louis region, we were behind other cities across the nation regarding other issues.

Unknown to me at the time was that my Mayor was grooming me to be his replacement. He had held a number of offices in the National League of Cities. He had been on the Advisory Council and the Board of Directors. One year he even ran for the president of the National League of Cities while I was with him. He introduced me to many people in the League as part of this grooming process.

One of the things the League officers did each year was to have a banquet meal away from the conference. My Mayor invited me to attend a few of these banquets with him. The one that is most prominent in my memory is the one at the Russian Embassy in Washington. The Embassy was located on a hilltop and I had been told by a League member that the Russians had antennas directed at the State Department and other selected government buildings and offices to intercept conversations. That made sense to me since I was aware that the Russians had essentially bugged the entire U.S. Embassy that they built for us in Moscow. They had even developed a passive bug that our best electronic equipment could not find for years.

In order for us to go to the Russian Embassy, the Embassy had to be provided the names of persons, several weeks in advance, that would be going to the banquet. I supposed that it was for screening purposes. I wondered about this. Was my name on their list of suspected spies since I had been in ASA? I figured that any KGB agent worth his salt would figure out my background in ASA. Would they be watching me the night I was there? I thought so.

On the night of the banquet we boarded two buses at the Washington Hilton a few feet from where President Reagan had been shot a few years before. When we arrived at the Russian Embassy there were barriers we had to navigate through. During the wait I looked at the Embassy but I didn't see anything that looked like directional antennas. Guards checked out ID's against the list of names they had been given by the League staff. There had been a cancelation by a person from the St. Louis area because of a broken arm and another council member from that city wanted to go in her place. No deal said the Russians, or was it just *Nyet*.

The Embassy atrium was furnished in beautiful white marble. The stairs from the ground level to the next level in the atrium were a partial spiral with black iron railings. At the top of the stairs were several tables with waiters, probably KGB I thought, pouring glasses of vodka and other liquors. There were several flavors of vodka to choose from. Who would have guessed there would be vodka at the Russian Embassy? Not being a vodka drinker I was pleasantly surprised at how good it tasted. I liked the lemon flavor the best. I slugged that stuff down way too fast. The glasses seemed to be so small.

For all of the time I was in the Embassy, I wondered if the waiters were wired. I was careful about what I said. I was sure one of the waiters was watching me because I was watching all of them. I also thought there must be electronic bugs everywhere in the building. At the same time I was pretty smug about being in the Russian Embassy. Hey, not every ASA spy gets into the den of the Bear.

After about an hour we were escorted into a huge banquet room and were seated. I remember that the meal was good but I have no memory of what we were served. I was relieved to leave that place and go back to the hotel and relax.

Had I been over-reacting or acting paranoid? I remembered my final debriefing in ASA and the list of countries I could never go to and Russia was on that list. Was I really that valuable to the communists since I left the ASA nearly forty years before? A better explanation may be that I was arrogant and felt more important than my being in ASA or of being a Council member meant to the Russians. Having even a little power can corrupt the mind.

Two or three years later the annual banquet was held at the Chinese Embassy in Washington. I was again asked to go. This time, I emphatically said no. There was no doubt in my mind that the Communist Chinese knew that I had been in Vietnam and in the ASA and we were still in a ceasefire with them and the North Koreans.

After eight years I decided I had done my civic duty and did not run for re-election. It was a privilege to serve the nearly thirteen thousand people I represented. It was a small thing compared to serving the people of America in the U. S. Army and ASA. I received a glass plaque from the Council and a brass plaque from the Fire Department thanking me for my service.

I don't have any more calls at night about seventeen cats running in and out of a house with broken basement windows and mating all over the neighborhood and *what am I going to do about it?* It's a house with cats, not a cat house! The police cars that used to bring my Council papers on Friday nights, now just go by on the street without stopping. All is well.

It took me three years to finally bag a trophy class black bear with a bow and arrow. I keep the mounted head in the basement and look at it occasionally as a reminder of earlier times. I hunt deer now. They don't climb trees to come after me.

They Came to Vietnam
By Randolph Willard

They came to Vietnam from Hollywood. One went to the North; one went to the South. One gave comfort and aid to the enemy; one gave comfort and aid to American soldiers, sailors, airmen and marines. One held America's service men and women in contempt with a hardened heart; one reached out to them with open arms and a loving heart. One never went into harm's way; one was shot down twice and had a bounty placed on her head.

One was born to Hollywood royalty; one was not. One was honored with golden statuettes by the Hollywood elite; one was not. One was not called to task by an America in turmoil; one was never truly honored for her sacrifice and love. One is despised and loathed by the Vietnam Veteran; one will be cherished forever.

There is a poster in a scrapbook that is cherished. It is forty-four years old. It is a photograph of a young actress posed in a golden bikini. She is blond and beautiful. To a lonely service man in a faraway hostile environment for the first time in his life, she is the epitome of the dream girl. Her name is Sandee.

Two actresses went to Vietnam. One sowed discord and discontent while avoiding the scaffold. This one received her thirty pieces of silver in fame and fortune and recognition by her Communist-Leftist friends. Chris Noel spread smiles and joy while avoiding assassination. Chris Noel spent her life in service to Veterans. Her sacrifice, suffering and contribution cannot be measured in gold.

I have witnessed our Presidents hand out the American Freedom Medal; many who received it for no other reason than sowing discord and creating havoc. I have witnessed where a perceived traitor received the American Film Institute award for her body of work. Where is Chris Noel when these honors are being handed out; not to be found?

Chris Noel may or may not have been a great actress. She may not even have been a good actress; but she was one hell of a caring human being. I don't expect Hollywood to honor her but American Vietnam Veteran's should and have. She was there for us, the American soldiers, sailors, marines and civilians that went to war when asked and did their job under adverse conditions. No war in American History received the damnation and condemnation from those who kept the home fires burning. Those men and women who served during this time were ill-treated, lied about and portrayed as crazoids just one nightmare from exploding into violent uncontrollable monsters.

I was one of those soldiers laboring in the Army Security Agency in a thankless task of reading holes in strips of paper. My job, as was a thousand others, was to sit in manmade caves and ferry messages by teletype. We shot no one, saw no action but we did our part. Our hours were long, the tasks tedious. My favorite detail was called the burn detail. We burned countless strips of paper and documents in an outdoor furnace on the grounds of Davis Station, the home of the 509th RRCUV. It was outside work. Most of the workdays were spent in what amounted to a cave. We often labored twelve hours on and twelve hours off. Free time was spent drinking, gambling, running whores and hanging out on Tu Do Street, the Saigon zoo or just wandering the giant Tan Son Nhut Air base. I liked to ride around in those three-wheeled motorized cyclos. I spent my time taking countless photographs, purchasing post cards and reading anything I could get my hands on.

I also listened to the radio. One of my favorites was actress, singer and disc jockey Chris Noel. Her show was called 'A Date with Chris' and she called us Luv. Much has been written about Chris Noel and many hold her dear. She did so much more than she is credited for by our liberal and anti-Vietnam media. I am amazed how this talented and devoted American is forgotten and virtually unknown today by so many.

Chris Noel did her part. She was legend. Her looks alone sent many into visions of the imagination that still are wonderful to think about.

She was born Sandee Louise Noel on July 2, 1941. She followed the usual path to semi-stardom. She was into batons, modeling and acting. She found herself running with and dating Hollywood celebrities. Her first movie was with Steve McQueen and Jackie Gleason. She acted with Elvis, Dennis Hopper and many more. She appeared in movies that show her blond, bikini good looks.

One day in 1965, she found herself in a Veteran's ward with Sandy Koufax, Governor Brown, Beverly Adams, Rowan and Martin, Eileen O'Neil, Ruth Berle and Chu Chu Collins. This hospital tour changed her destiny. Her friend Jack Jones got her an interview with Armed Forces Radio and Television. The rest is history, as Chris Noel would soon become the darling of G.I.'s in Vietnam and other parts of the world who thought she was talking to them as she recorded her show, a record that was to play all over the world. During her show, they placed tidbits from Earl Nightingale.

She toured Vietnam and went to all the backwaters and outposts that few feared to tread unless their military commanders ordered them there. She was in helicopters that crashed, near enemy action, and through it all, she brought a smile, a hug and a fantasy to thousands of American soldiers, sailors and marines in a hostile country. A bounty was placed on her head. The enemy thinking her death by enemy action would wreak havoc on American morale.

She was often seen in jungle fatigues, but just as often would appear wearing a short mini-skirt, and knee-high boots. I don't think it mattered much to the G.I. fighting for a glimpse of the voice he was listening to on the radio.

I never met or even saw Chris Noel in person. She was a voice on the radio. I did not even know she was an actress. My youth in the beach party era was spent at one grind mill job after another. I found a poster of her in a golden metallic-like bikini on one of my trips to Cholon and hung it up in my locker, but was careful to take it down if an inspection was imminent. I had a picture struck down from my locker door by an irate Colonel. He used a swagger stick too and my sergeant rolled his eyes.

I did my tour, went to the Bad Aibling Field Station in West Germany. I sent my photo albums and souvenirs home to my mother's house. I never thought much about Chris. I found other women to fall asleep thinking about. I even found a live one to fall asleep with me.

VFW magazines did a series on music in the military. I remembered the great sixties sound track from Forest Gump. I saw no mention of Chris Noel. I found my scrapbooks and found the several I made while at Davis Station. The poster of Chris was neatly folded into the three sections. I showed it to my son who asked who she was. Who is she indeed? I could not find a copy of 'Matter of Survival' that I could afford but I did obtain a copy of 'Vietnam and Me'. I bought a copy of Chris Noel; Confessions of a Pin Up Girl'.

I started asking around. No one seemed to remember Chris. I thought that this was sad since that pig of an actress had received so much acclaim and recognition since her trip into the enemy stronghold where she posed for pictures on active weaponry being used to kill Americans at the time. To this day, it still boggles my imagination that she was not hung. I refuse to see any film in which Fonda appears.

While writing and researching this story, I found that many still remember Chris. I found she helps Veterans to this very day and that she gives speeches appearing at Veteran's conventions and reunions. I found that she has been honored. She is the recipient of the Distinguished Vietnam Veteran Award from the Veterans Network in 1984. I also found out that our own government has not taken note of her or honored her. We have stamps for cartoon characters and folks I would absent myself from in an effort to avoid them. It would be nice if she got a medal of some sort while she can acknowledge it.

She served her country most nobly when many of her profession were calling for its downfall. Chris suffered turmoil in her life. She had what is now recognized as PTSD. She lost her Green Beret to suicide because of his PTSD. Her career faltered and fell as she sought for the answers that all veterans seek. I won't dwell into her failings. I want to rejoice in her successes. She joined our side when so many were fleeing.

Americans are good at forgetting its heroes. I know folks who cannot tell me who Audie Murphy or Roy Rogers were. I remember and I do my best that others remember too.

Others went to Vietnam. Some engaged in fierce battles and suffered atrocity. Some spent more time in an enemy's prison than they did as warriors. Many won medals for valor and sacrifice. Many toiled as anonymous clogs in a giant military wheel. I was one of these men. The only medals and honors I have are the showing up kind. I showed up. I followed orders. My military job was mundane and repetitive. I am not surprised that so many found solace in drink, and other pursuits.

I don't know what it was about the Vietnam War and why so many Americans did not want to support the military. I cannot speak of others, but working in the Army Security Agency gave me a sense of pride and accomplishment. I was a high school graduate and I was working with college men, one who even had his PhD in chemistry. The men I worked with, it was a time when men worked in male units, were fine fellows with hopes and dreams of what they wanted to accomplish either in the service or a subsequent civilian life. I was exceedingly pleased that I held a Top Secret Crypto clearance. No one I knew back home ever had such an honor.

I came to the agency by happenstance. I tried to enlist and was sent back home as unfit. Later, I received notice telling me to report. I went to the board and was told that I still had to go take a second physical. This meant a bus ride to Montgomery, Alabama. They told me I passed after sending me to an outside optometrist. I was underweight but the private doing the weigh-in gave me eight pounds. I knew another man at the physical and he said he was going ASA. I knew about ASA as an older man on my street had been in ASA. He made a career of the army and stayed with the agency after it merged in the mid-seventies. My goal was to be a writer and the recruiter told me that ASA had a journalism school. The downside was it was not available. I signed up anyway and boarded a jet plane to Columbia, South Carolina where I did basic training at Tank Hill in A-1-1. I remember still how those smiling sergeants turned into growling fiends once I took that oath.

I was no longer in control of my own life. The army owned me. I took test after test and somehow ended up at Fort Gordon in Augusta, Georgia. To this day, I think it was because I could type. I sped through the typing week the first day. The rest of the week I raked dirt. I did get moved up a week.

I remember filling out the dream sheets. I put down Vietnam. Since I was in the Army and the war was in Vietnam then that was a good place for an aspiring adventurer. As far as I know we all ended up in Vietnam. My next stop was the Hotel Saint George in Saigon. Everyone wanted to stay in the Saigon area. I put down no preference, as did my friend Bill Smith. We both got Davis Station and the others went out to the boonies.

I never saw any gunplay but I saw fighting men. I kept hearing the men on my trick damning the place and pining for a girl or wife back in the world. Personally, this was no issue for me as no one was waiting on my return. I had two pen pals, one a cousin and one a friend. I don't count letters from my Mom or brother. Although at mail call it is always nice to have one's name called. Nothing is as sad as being eager then having to walk away empty handed.

After Vietnam, I had a tour in Germany and one in Japan. I came back to the States and to my old grind mill job at the Pensacola News Journal. I regrouped for a few years and attended a local junior college. I moved to Georgia and became a Peace Officer and I followed this trade for the next thirty-five years. I met my wife at a police department. I liked her so much I have managed to keep her. It is my one great accomplishment.

I have thought how the army and particular ASA benefitted me in the life I led later.

Coming out of an impoverished background where teachers, bosses and others did not give me much credit or any chance at success into an elite organization such as the Army Security Agency, did a world of good for my self-image and self-respect. I liked writing down my having the aforementioned Top Secret Crypto clearance. It was hard describing exactly what ASA did. People thought I was either a security guard or a spy. Whatever confidential information I knew is long forgotten. Most of the time I was too busy getting the job done to worry about what the generals and officers were doing. I was more worried about what my sergeants were up to because if a name was called it meant extra work.

I managed to stay out of trouble. I did receive the Good Conduct Medal. I qualified Expert with the M-16. I always thought I did well on this because I had never owned a rifle and the M-16 confounded the hunters in my basic unit. I also scored 475 out of 500 on my final PT exam. I owe this good showing to the fact that I spent most late afternoons working out when others more fit than I were given the rest of the day off. I was in what was called the Goon Platoon. I and other physically out of shape men did calisthenics in a sawdust pit. We also ran until we threw up or collapsed on the ground. This added to my confidence.

Years later when I was at the FBI Academy, I was part of a group that ran, after classes were out. My group was called the grey panthers because we were the oldest of the attendees. Every one of my group finished the entire extra hours training without dropping out when many in the younger and faster group dropped out. The funny thing is that one could move in and out of the running groups at will. I guess egos played a hand. I kept thinking about the old Goon Platoon and I just kept going on those long runs through the back roads at Quantico Marine Base.

At one time or another I worked in all areas in law enforcement. I attribute my military training and experiences as giving me an advantage. I knew how hard I could be pushed and I learned that complaining only annoys those who make your career path decisions for you.

There were thousands like me in Vietnam. We charged no hills, stormed no beaches, and killed no one with a rifle or hand to hand. We did jobs that had to be done. Often our work site was in awful places in terrible living conditions. We were summoned and we went. We did not skate out. We showed up for work, did our job and came home to live our lives. I would have liked to have won honor and glory on a combat field. I wanted to know if I would measure up or cry like a baby. None of this was for me. I did what they asked, as did countless others.

No one asked Chris Noel and Jane Fonda to go to Vietnam. They went on their own volitions. Each had a different mission. One came to help and bring cheer and boost the morale of lonely and oftentimes frightened men, risking her own life in the process. One came to sow discord and disharmony, giving her country a black eye and undeserved fictional reputation carried by the media and Hollywood as fact.

I don't know Chris Noel or Jane Fonda from Adam's cat but if I had to spend time with one on a stuck elevator or life raft, I hope it is Chris Noel.

Chris performed with the legendary Bob Hope. Hope's theme music was 'Thanks for the Memories'. From one Vietnam Veteran to another, 'Thanks for the memories, Chris Noel.'

Looking Back Over a Half Century
By John *Mad Jack* Klawitter

It was autumn of 1962. I was 24 years old and had been a student ever since I was five or six. I had an uncle who pounded into me that knowledge was power, and a schoolteacher mother who believed ignorance was a sin, and after those formative years I had earned a B.A. in English and a B.S. in Geology, and had minors in Religion, Philosophy, and History.

At the time, I was studying Graduate English on a fellowship at UCLA, so naturally I was pissed when my radicalized, black-as-coal Jamaican girlfriend informed me I was a dumb-as-a-brick farm boy who didn't know shit anything about anything.

She had conned me into showing up for a rally in front of Royce Hall. She gave me a hurried peck on the cheek and shoved a handful of pamphlets in my hand. They featured a crude sketch of a tank running over some cone-hatted peasants and a headline "Invader-Gangster American Colonialists—Get out of Vietnam!"

"Hand these out, Jack," she said. "Maybe you'll learn something!"

As you might expect, my mood turned sour fast. I kicked around, feeling used and cheated. I'd thought we were going to some sort of pep rally for a big football game, and I was entertaining wild fantasies there would be crazed chanting around a big bonfire followed by hot sex at her place. But that was not to be.

Apparently, a few years before, President Eisenhower had sent the French in Indochina a few hundred war surplus tanks, and a few hundred military advisors, and the folks who'd put out this brochure didn't like that gesture. I was really miffed.

This business of some mysterious third party protesting an arcane problem between our old WWII allies and some guy named Ho Chi Minh that nobody even knew about didn't set right with me so I shouted back at her.

"Well, you don't know anything, either!"

"Jack, I stood right there while the policeman shot my sister in the face!"

"Where was that, in Saigon?"

"No, in Kingston, you dumb shit! But it's all the same"

"It's not all the same! You've never been to Vietnam. Yet you're handing out pamphlets condemning our country."

"You are the absolute dumbest, stupidest person I have ever met!"

"At least I don't sprinkle sand in front of my door to keep bad witches out!"

She gave me *the look of fury.* I figured I'd be sleeping in my own apartment that night anyway, so what did it matter?

"You dumb island girl! You're just a tool! You're being used! You have no idea what you're talking about!"

"You never been there either, farm boy! You never been anywhere!"

I was way beyond rational thought.

"Okay, I'll tell you what. I'll go and see what it's all about. Then I'll come back and tell you about it!"

She laughed. It was an ugly barking sound, her memories of island atrocities whether real or imagined spilling out of her.

"Yeah. You do that, Jack."

Well, over the next week or so it still seemed like a good idea. I didn't trust her because I believed she wanted us to make a baby and get hitched, in that order. I wasn't getting anywhere with Graduate English, which wasn't at all about creative writing like I'd imagined. And Hemingway had gone and seen his war. It seemed to me that was the way to do it. I got in my battered old car and drove from Westwood Village to Santa Monica where an army recruiter chewed on a hamburger while he agreed with everything I said.

After the army I never did get back to hash things out with my girlfriend. In fact, I never saw her again. But I had gone and seen for myself the place where the anti-war crowd said our tanks were crushing the poor cone-hat peasants.

I found out first hand something about what was going on in that part of the world. And I came to believe that my girlfriend and a lot of the rest of us Americans were being propagandized by a loose group of people who were banded together by conviction. They hated the United States and just about everything we stood for. Don't discount the effectiveness of jealousy or hate, no matter how irrational or misplaced. It worked pretty well for them.

Over the war years, these people managed to convince many of us we were not noble or kind or righteously engaged in the conflict in Southeast Asia. Those of you old enough may remember we won that war on the distant battlefield, but we lost it back at home. We lost it in the press, on the radio, on the nightly television reports. And so, in reflection, seeing how it turned out, today I have to admit 50,000 of our best young troopers died in vain. If you are going to engage, stay the course, no matter how bitter. Check out George Washington's winter time trials at Valley Forge.

And, if you still don't know how you feel about this, the next time you're in D.C. remember what I'm writing here as you walk slowly down slope into the crushingly sad Vietnam War Memorial. Think it through yet one more time. Remember what it was like for them, for us, for you. Remember how the press and the so-called *intelligentsia* turned against us. The pigs-blood balloons. The draft card burnings. The craven Weathermen who got away with bombs and murder. The water-in-the-knees injections from obliging family doctors. The indignant priest-agitators. The student rioters with their bull-horns. The cowardly exits to Canada and Sweden.

I went west across the Pacific over a half century ago, as did many of you. It was far from a perfect, noble war. If you've read Hemingway, Conrad, or Heller – or James Webb, Robert Stone or Michael Herr – you know they never are. And if you lost a brother, son or father you know the Nam conflict wasn't worth it to you personally. And no, we are not a perfect nation. But we went to Vietnam to fight a brushfire war against Communism for John F. Kennedy. We went to save a people. To free a people, to help a people enjoy the sort of life we have. We lost our heart, our spirit, our courage…and over a million Vietnamese were slaughtered after we pulled out in disgrace. But never doubt that we were and are an exceptional nation, a country willing to sacrifice for abstract and yet very real principles…life, liberty, human dignity. Not perfect, no. But exceptional, yes.

I think, if you haven't read it, Bernard Fall's book of essays, *Last Reflections on a War* (published in the 1970s) still holds up after all these years.

Nonetheless, full of piss and vinegar, I joined the U.S. Army in autumn of 1962. After Basic Training, I took a 47 weeks intensive Vietnamese course at Monterey Language School. I took courses in Oriental politics, history and culture at the Monterey Institute of Foreign Studies. I spent six months at the Puzzle Palace at Fort Meade learning decoding and crypto. I was in Saigon from June, 1964, to August, 1965, decoding covert Viet

Cong and Viet Minh messages at an Army Security Agency outpost located at Tan Son Nhut that some of us nicknamed The White Shack. (There is a picture of the entrance on the front cover of TANS III, ably edited by Wayne Munkel) And I had a Top Secret clearance. In short, call me crazy but I'd done my best to fulfill my promise to find out about what was going on with the cone-hatted people and why we were running them over with tanks.

I came back stateside ready for arguments, on the prowl for dupers I could take down with what I'd learned first-hand. And that's when I learned an even bigger lesson. The people who said we were evil colonial gangsters didn't want to hear it. They had already made up their minds a long time ago that we were the bad guys, period. They didn't want to hear anything from people like me who'd actually looked and listened and experienced what was there and had some different ideas based on a combination of intellectual pursuit and actual experience.

Yes, bad enough...but there's worse news: The folks who duped my Jamaican gal pal fifty years ago are still around. Some of them are actually still alive and actively trying to pull down the US, and others who were their students have become just as nimble at hating this country as were their teachers. In fact, the movement is stronger than ever before. You find it on progressive talk shows, in entertainment magazines and on social media – sincere believers tearing down traditional American values, laughing at the way we are. They're in the justice system gnawing away at the constitution. They're in government jobs, working to stifle free enterprise and present a more controlled distribution of wealth system. They're in colleges and universities teaching our kids and grandkids many things other than what a great and good nation this is. For reference, see

DUPES, by professor Paul Kengor, should you think your own *Mad Jack* is simply one lone madman howling in the dark. I appreciate that book for its carefully researched historical backing and footnotes. Try it; you might like it.

Alright, what can we do? What can any one person do? I don't have a snappy, easy answer for that one. I guess, stay the course. Be strong in your convictions. Be skeptical when you hear people who should know better dismiss what is special about the United States of America. Stand up for our values, our system of government, our enterprising way of life, our freedoms of belief and expression and equality. Believe in this country, or don't. Lots, you'll see around you, don't. I still do, and I firmly believe that so do many of you.

Living with Terrorism
Wayne Munkel

I arrived in Vietnam April 6, 1963. I did not know then that the VC had begun urban attacks on Americans in February, 1963. I became aware of attacks when I would venture into Saigon at night and hit the bar scene. I heard numerous explosions in different parts of the city. I would read about the bombings in the newspaper within a few days, that a bar frequented by Americans had been blown up. One such bar was a block outside the main gate of Tan San Nhut where I lived. I went by it for weeks afterward. The whole front of the bar was blown off.

On February 9, 1964, two of the men of my ASA unit, Arthur Wayne Glover and Donald Taylor, were killed by a bomb planted under the bleachers at a softball diamond on Tan San Nhut where I lived. One of the men on my shift, Arthur Wayne Glover was killed. He had just pitched the first game of a double header and sat down as the second game began. He was struck with five pieces of shrapnel one of which pierced his heart. He lived for about thirty minutes and never made it to medical help. Another ASA soldier, Donald Taylor, was nearly blown in half and died instantly in the explosion. I attended the ceremony that sent them on their way back to the U. S. and their final resting places.

The Capital-Kinh Do Theater was a theater that was exclusive to Americans. I attended the Capital-Kinh Do Theater and saw the movie '*The Longest Day*' about the D-Day landings. The movie was in French with English sub-titles. The theater was bombed on February 16, 1964. Three Americans and nine Vietnamese were killed.

The USS Card was sunk May 2, 1964, by the VC in downtown Saigon. The ship was raised and went to the Philippines for repairs.

Shortly after I left Vietnam, the Caravelle Hotel was bombed on August 25, 1964. I had a few drinks in the rooftop bar at this hotel and had been there several times.

Terrorism is not a one-time event. It is a series of sudden, calculated attacks aimed at creating a state of fear and chaos in the population attacked. It is mass murder. Terrorists are in it for the long haul and are patient and creative.

One of my best buddies was in Saigon and witnessed a person riding a bicycle when it exploded. Body parts flew in all directions and bounced off the walls of buildings on both sides of the street. Bicycle bombs were commonly used to blow the fronts of bars into the bar. The bicycle was placed against the bar and set off when Americans were inside. American deaths were prevented by the premature explosion that killed this VC terrorist. Terrorists stop when they are dead.

I became very aware of the danger of terrorist attacks after hearing the numerous explosions and reading about Americans being killed by the bombs. I was never near one of those explosions but I didn't have to be to be affected by them. During my nearly seventeen months in Vietnam Americans were being killed in increasing numbers. As the time of my departure neared, I became more fearful and guarded about where I went around Saigon. It felt like I had a bull's eye on my back when I was in Saigon. After living with fear and anxiety about being killed there by terrorists, I resolved to myself that I would never again be frightened by terrorists. Little did I know this resolve would be challenged in the years ahead.

We were looking forward to our son's wedding in Sedona in September. Our oldest daughter was flying to Phoenix from Florida. Our youngest daughter was flying from New York City. We were all going to meet in the airport in Phoenix and then go to the Grand Canyon before arriving in Sedona.

I was getting ready for work as usual on September 11, 2001 when my oldest daughter called and said, "Dad, turn on the TV. We are under attack!"

And so we were. Al Qaeda had declared war on the U.S. years before and had finally struck us.

Unknown to us was that our youngest daughter was on her way to work in New York. She was on a subway train from Queens where she lived to Cherry Lane Theater in south Manhattan where she was working on a play. The train line went through the World Trade Center station. The train stopped at 42nd Street because the first tower had collapsed. She began to walk toward the theater and saw the second tower fall. At one point she got in line to give blood at a south Manhattan hospital. After an hour she moved on and saw a friend pouring coffee for first responders. She spent the next several days on Manhattan because public transportation was shut down. It was a difficult time for her.

After five days airlines were allowed to fly again. Until that happened we were not sure we would be at our son's wedding. My wife and I left St. Louis early on the day of our scheduled flight. There were only about thirty people on the flight because many people had canceled out of fear of terrorism. When our youngest daughter arrived in Phoenix we spent more than two hours in an airport restaurant debriefing her and listening to all she had been through during the previous week. Some things were hard to hear. Our 'baby' had been so close to the terrorist attack and was unhurt physically but was very stressed by her experiences. Once again, like those days in Vietnam, terrorists had touched me.

Our older daughter arrived late in the afternoon and we were off to spend the night with Jack and Becky Waer. I was unloading our rental car at our hotel in Sedona a couple of days later when my sister and her husband came toward our room. I could tell by their posture that something was amiss. They announced that our brother's wife had been killed in a car crash as she was getting her dress for the wedding. Our brother would not be coming to the wedding; he was planning a funeral. The next two days were a roller coaster of emotions for us.

So, what do I do now with terrorism rampant in the world and on the streets of America? I keep the perspective that I am not going to be constantly afraid. That would mean that terrorists are successful in terrorizing me. I am vigilant and keep abreast of what is happening in the world and in certain countries where terrorism is being exported. I am amazed at how ignorant many Americans are of the threats directed at America and for those who are aware, the lack of concern about the threat of terrorism. ISIS has declared war on America. What towers will fall when they attack?

ABOUT THE TANS CONTEST

TANS started as a short story contest. It still is a short story contest. If you have the qualifications, you can enter it. And if you have the write stuff, you might be awarded a fancy piece of paper saying your story is a marvel and even a winner. The stories must be remembrances, a category of tale-telling that the literary scholars of our time call Creative Non-fiction. This is a fancy way of saying you are writing from your true experiences. More specifically, to qualify these stories have to be the remembered experiences of U.S. and allied military personnel related to their adventures, misadventures, triumphs and blunders in military Intelligence. We would probably also entertain the submission of engaging stories from the enemy, though to date none have been submitted. For further information please visit www.oldspooksandspies.org.

ABOUT THE TANS BOOKS

Stories submitted for the TANS Contest are eligible to be published in the TANS books, of which this is the fourth. They are selected by the TANS anthology editors with the generous input and advice of various members of the Executive Committee of the Old Spooks & Spies.

TANS I was published in 2002, edited by John *Mad Jack* Klawitter. TANS II was published in 2011, edited by Wayne Munkel. TANS III was published in 2014, also edited by Wayne Munkel. TANS IV, published in 2016, was edited by Wayne Munkel with assistance from *Mad Jack.*

TANS I, II, III and IV are all available on line from amazon.com and bn.com, and may be ordered from your local fine bookstore.

Stories submitted for the contest and published in the TANS anthologies remain the property of the individual authors, who grant rights to the Old Spooks & Spies for publication in the anthologies.

FIRST/LAST WORD

"TANS" is an acronym for the phrase "That Ain't No Sh*t", a common enough expression among military personnel sometimes used in clubs, bars, and at reunions and other gatherings, used to declare with the most emphatic emphasis that the words spoken, the stories being told, the claims no matter how amazing and outrageous, are in point of fact as true as true can be for soldiers with security clearances and a fond regard for their military brothers and sisters who served in troubled and often perilous times. In some few instances names are changed to protect both the innocent and the guilty, and exact times and dates may wander due to faulty memories. Where possible, these have been noted, but, after all, these are old war stories. If you are a protestor or a shirker and wish to complain, please remember we were there and you weren't.

-mj